Second Order Differential Equations

Gerhard Kristensson

Second Order Differential Equations

Special Functions and Their Classification

 Springer

Gerhard Kristensson
Department of Electrical
and Information Technology
Lund University
SE-221 00 Lund
Sweden
Gerhard.Kristensson@eit.lth.se

ISBN 978-1-4939-0177-7   ISBN 978-1-4419-7020-6 (eBook)
DOI 10.1007/978-1-4419-7020-6
Springer New York Dordrecht Heidelberg London

Mathematics Subject Classification (2010): 33XX, 33CXX, 34XX, 34MXX

Printed on acid-free paper

Springer is part of Springer Science+Business Media (www.springer.com)

To the Memory of my Father, Gunnar

Preface

With modern computers and software tools, there are, in general, no problems generating numerical values to various kinds of special functions. Indeed, excellent software, such as Mathematica, Maple, and MATLAB, can generate a large variety of special functions with high precision. Moreover, the general topic of special functions is well covered in the literature, see, e.g., [2, 3, 8, 14, 20, 28]. However, the knowledge of the overall structure and relationship between the different special functions is often lacking nowadays. This textbook tries to remedy that need. The presentation of the theory in the book does not deal with the particular properties of various special functions, but rather focuses on the generic connection between the functions and the families they belong to.

The way the special functions are introduced and classified is the subject of the current textbook. The aim is to provide a self-contained treatment of the subject intended for the upper undergraduate, graduate student, or the researcher in mathematical physics who has a need to understand the underlying systematics of special functions. There are many ways of approaching this subject indeed. The most popular ones are:

- Classification and systematics based upon the singular behavior of the coefficients of the underlying ordinary differential equation, see, e.g., [12, 23]
- Group theoretical approach, see, e.g., [17, 27, 29]
- Classification and systematics based upon integral averages, see, e.g., [4]

We pursue the first, most traditional, approach in this textbook. The singular behavior of the coefficients manifests itself by the number of singular points — poles or branch points — in the complex plane. For systems with less than three singular points, the solutions of the ordinary equation belong to a rather well defined class of functions. As the number of singular points becomes three or more, the solution class becomes "rich" in the sense that most functions encountered in mathematical physics are found among these solutions. The purpose of this book is to explore and understand the systematics of these classes of functions and their relations to the many special functions encountered in applications and in mathematical physics.

The focus is on the overall relationship of these solutions, rather than the analysis of the particular special functions themselves.

This textbook originates from a series of seminars in the mid-1970s on special functions given by the late Professor Nils Svartholm at the Institute of Theoretical Physics, Chalmers University of Technology, Göteborg, Sweden. Incidentally, Professor Svartholm made several important contributions to the solution of Heun's equations in the late 1930s, see Section 8.4. The general outline of Professor Svartholm's notes is to some extent kept, but numerous extensions have been made in order to make the text more complete.

The text intends to cover a three- to four-week upper undergraduate or graduate course on the subject. The prerequisites of the course are analytic function theory on the level of, e.g., E. Hille [13] or R. Greene and S. Krantz [9]. Specifically, the reader should have basic knowledge of the method of residues and multi-valued functions. To make the textbook more self-contained and complete, a series of appendices are found at the end of the book, which contain specific background material. At the end of each chapter, there are problems that illustrate the analysis in the chapter. It is recommended that students solve these problems in order to get a better understanding of the theory. Problems marked with a dagger, [†], indicate problems that are more difficult. A solution manual to all problems is available at the home page of the author.

I am most grateful to Professor Anders Melin, who has been very supportive and helpful during this whole project. He has contributed numerous valuable comments and improvements to the text. This is particularly true for Appendices A and B in which he has given esteemed input and criticism. Writing this book has taken most of my leisure time, and thanks to an understanding wife, the project had an happy ending. Thank you Mona-Lisa! The author is also grateful to Martin Norgren, Kristin Persson, Daniel Sjöberg, and Christian Sohl for finding typos. Finally I like to thank Springer, especially Vaishali Damle and Marcia Bunda, for very constructive collaboration.

Lund, May 5, 2010   *Gerhard Kristensson*

Contents

Chapter 1
Introduction

The question of how the equations of physics originate is adequate. They usually arise from a boundary value problem that models a particular physical problem of interest. Typically, a boundary value problem consists of

$$\begin{cases} \text{Partial differential equations} \\ \text{Boundary value and/or initial value conditions} \end{cases}$$

Sometimes the physical problem itself is modeled as an ordinary differential equation (ODE) — often of second order. Moreover, solving partial differential equations of various kinds introduces the ordinary differential equation by the methods through which these equations are solved. One striking example of how second order differential equations arise from a more general problem is the method of separation of variables. As an explicit example, we take the Helmholtz equation in three dimensions, which reads

$$\nabla^2 \Phi(x,y,z) + k^2 \Phi(x,y,z) = 0$$

The ansatz

$$\Phi(x,y,z) = X_\alpha(x)Y_\beta(y)Z_\gamma(z)$$

where (α, β, γ) belongs to an index set I, implies that each function $X_\alpha(x)$, $Y_\beta(y)$, and $Z_\gamma(z)$ satisfies a second order ODE. Requiring some additional conditions on these functions, the complete solution of the Helmholtz equation is obtained as a series

$$\Phi(x,y,z) = \sum_{(\alpha,\beta,\gamma)\in I} X_\alpha(x)Y_\beta(y)Z_\gamma(z)$$

Specifically, the separation of variables is successful in 11 coordinate systems, e.g., the spherical coordinate system, (r, θ, ϕ), leads to a set of ordinary differential equations

G. Kristensson, *Second Order Differential Equations: Special Functions and Their Classification*, DOI 10.1007/978-1-4419-7020-6_1,
© Springer Science+Business Media, LLC 2010

$$\begin{cases} \left(\dfrac{d^2}{dr^2} + \dfrac{2}{r}\dfrac{d}{dr} - \dfrac{l(l+1)}{r^2} + k^2 \right) R_l(r) = 0 \\[2ex] \left(\dfrac{d^2}{d\theta^2} + \cot\theta \dfrac{d}{d\theta} - \dfrac{m^2}{\sin^2\theta} + l(l+1) \right) \Theta_{lm}(\theta) = 0 \\[2ex] \left(\dfrac{d^2}{d\phi^2} + m^2 \right) \Phi_m(\phi) = 0 \end{cases}$$

which all can be written as differential equations with coefficients as rational functions (change the variables $\theta \rightarrow u = \cos\theta$).

In all the 11 separable cases, the method of separation of variables leads to ordinary differential equations of second order where the coefficients are rational functions[1] [3, 5]. It is therefore of some interest to study ordinary differential equations of the following form:

$$A(x)\frac{d^2u(x)}{dx^2} + B(x)\frac{du(x)}{dx} + C(x)u(x) = 0$$

where $A(x)$, $B(x)$, and $C(x)$ are polynomials (or rational functions) of $x \in \mathbb{R}$.

We extend the independent variable to the complex plane $\mathbb{C}$ and let $x \rightarrow z = x + iy \in \mathbb{C}$. The complex roots of the leading function $A(z)$ are vital. We write $A(z)$ as

$$A(z) = \prod_{i=1}^{m}(z - a_i)$$

and classify the solution depending on[2] $m \in \mathbb{Z}_+$. In the current textbook, ordinary differential equations and their solutions with at most four roots, $m = 1, 2, 3, 4$, are analyzed. The appropriate form therefore is

$$A(z)\frac{d^2u(z)}{dz^2} + B(z)\frac{du(z)}{dz} + C(z)u(z) = 0 \tag{1.1}$$

where $A(z)$ has the form above, and $B(z)$ and $C(z)$ are meromorphic[3] or rational functions in a domain $S \subset \mathbb{C}$. Most special functions that are encountered in mathematical physics are included as a solution of an ordinary differential equation of this type.

---

[1] A rational function is defined as a quotient $p(x)/q(x)$ between two polynomials $p(x)$ and $q(x)$, where $q(x)$ is not identically zero.

[2] The notation used in this textbook is presented in Appendix F on page 213.

[3] A meromorphic function is defined as a quotient $p(z)/q(z)$ of two analytic (holomorphic) functions $p(z)$ and $q(z)$, where $q(z)$ is not identically zero.

Chapter 2
Basic properties of the solutions

In this chapter, the basic concepts of second order linear differential equations are introduced, such as regular and singular points — both located in the finite complex plane or at infinity. Moreover, the general properties of the solutions at such points are analyzed.

2.1 ODE of second order

2.1.1 Standard forms

The **standard form** of a second order linear differential equation (ODE) is, cf. (1.1)

$$\frac{d^2 u(z)}{dz^2} + p(z)\frac{du(z)}{dz} + q(z)u(z) = 0 \tag{2.1}$$

where $p(z)$ and $q(z)$ are analytic functions in a domain $S \subset \mathbb{C}$, or analytic in S except at a finite number of isolated points, i.e., meromorphic functions in S. The domain S can be the entire complex plane including the ∞-point — the extended complex z-plane.[1] We seek solutions $u(z)$ to this equation that are analytic in at least some parts of the domain S.

The equation can be transformed to a **reduced form**

$$\frac{d^2 u_1(z)}{dz^2} + q_1(z)u_1(z) = 0 \tag{2.2}$$

with the following change of dependent variable[2]

---

[1] A function $f(z)$ is analytic at infinity, $z = \infty$, provided the function $f(1/\zeta)$ is analytic at $\zeta = 0$. The meromorphic properties at infinity are defined in the same way.

[2] If the domain S is not simply connected, multi-valued functions occur, e.g., if $S = \mathbb{C} \setminus \{0\}$ and $p(z) = c/z$. To avoid this situation, assume S is simply connected.

G. Kristensson, *Second Order Differential Equations: Special Functions and Their Classification*, DOI 10.1007/978-1-4419-7020-6_2,
© Springer Science+Business Media, LLC 2010

$$\begin{cases} u(z) = u_1(z)\exp\left\{-\frac{1}{2}\int_b^z p(z')\,dz'\right\} \\ q_1(z) = q(z) - \frac{1}{2}\frac{dp(z)}{dz} - \frac{1}{4}(p(z))^2 \end{cases}$$

where $b \in S$. It is easy to prove that if $u_1(z)$ is a solution to (2.2), then $u(z)$ is a solution to (2.1), and, conversely, if $u(z)$ is a solution to (2.1), then $u_1(z)$ is a solution to (2.2), see Problem 2.2.

In fact, the reduced form of the differential equation, (2.2), is a special case of a more general transformation of variables, given by the following theorem:

Theorem 2.1. *Let $u(z)$ be a solution of*

$$\frac{d^2u(z)}{dz^2} + p(z)\frac{du(z)}{dz} + q(z)u(z) = 0$$

where $p(z)$ and $q(z)$ are analytic in a domain S.

a) If $z = g(t)$, where $g(t)$ is analytic in a domain T, such that the image is contained in S, i.e., $g(T) \subset S$, $g'(t) \neq 0$, $t \in T$, then $v(t) = u(g(t))$ satisfies

$$\frac{d^2v(t)}{dt^2} + \left(p(g(t))g'(t) - \frac{g''(t)}{g'(t)}\right)\frac{dv(t)}{dt} + q(g(t))\left(g'(t)\right)^2 v(t) = 0$$

b) If the function $h(z)$ is analytic and nonzero in S, then $u(z) = h(z)v(z)$ defines a function $v(z)$, which satisfies

$$\frac{d^2v(z)}{dz^2} + \{p(z) + 2f(z)\}\frac{dv(z)}{dz} + \left\{(f(z))^2 + f'(z) + p(z)f(z) + q(z)\right\}v(z) = 0$$

where $f(z) = h'(z)/h(z)$.

This theorem is straightforward to prove and left as an exercise, see Problem 2.1.

2.1.2 Classification of points

A point $z = c$ in the finite complex plane is classified depending on the analytic properties of the functions $p(z)$ and $q(z)$ at $z = c$. The following definition is essential for the analysis of the ordinary differential equations treated in this book.

Definition 2.1. A point $z = c \in \mathbb{C}$ ($|c| < \infty$) is called a **regular point** of the differential equation (2.1) if $p(z)$ and $q(z)$ are analytic in a neighborhood of $z = c$. A point $z = c \in \mathbb{C}$ ($|c| < \infty$) is called a **singular point** of the differential equation (2.1) if $p(z)$ or $q(z)$ have a singularity at $z = c$.

Similarly, a point $z = c \in \mathbb{C}$ ($|c| < \infty$) to the reduced form (2.2) is classified as a singular or a regular point depending on whether $q_1(z)$ is singular or analytic (regular) at $z = c$, respectively.

The point at infinity is special, and the classification of this point as a regular or a singular point is postponed to Section 2.5.

2.2 The Wronskian

We start with a definition and a lemma.

Definition 2.2. Let $u_1(z)$ and $u_2(z)$ be two meromorphic functions in a domain S. The Wronskian[3] of the functions $u_1(z)$ and $u_2(z)$ is then defined as

$$W(u_1, u_2; z) = u_1(z)u_2'(z) - u_1'(z)u_2(z), \quad z \in S$$

From this definition, the Wronskian depends on the functions $u_1(z)$ and $u_2(z)$. The following lemma shows that if $u_1(z)$ and $u_2(z)$ are solutions to (2.1), $W(z)$ depends only on the variable z and the function $p(z)$.

Lemma 2.1. *Let S be a connected domain S in the complex plane containing only regular points of the differential equation (2.1). Then the Wronskian of two solutions, $u_1(z)$ and $u_2(z)$, to (2.1), i.e.,*

$$W(z) = u_1(z)u_2'(z) - u_1'(z)u_2(z)$$

in S satisfies

$$W(z) = W(a)\exp\left\{-\int_a^z p(z')\,dz'\right\}$$

where a and z are points in S.

Proof. Derivation and use of the differential equation give

$$\frac{dW(z)}{dz} = u_1(z)u_2''(z) - u_1''(z)u_2(z)$$
$$= -u_1(z)\left[p(z)u_2'(z) + q(z)u_2(z)\right] + \left[p(z)u_1'(z) + q(z)u_1(z)\right]u_2(z)$$
$$= p(z)\left(u_1'(z)u_2(z) - u_1(z)u_2'(z)\right) = -p(z)W(z)$$

The Wronskian therefore becomes[4] (a and z are regular points of the differential equation (2.1) in S)

$$W(z) = W(a)\exp\left\{-\int_a^z p(z')\,dz'\right\}$$

by integration. □

---

[3] Józef Maria Hoëne-Wroński (1778–1853), Polish mathematician.
[4] See also the comment made in footnote 2 on page 3.

From the result in this lemma, we see that if the Wronskian of two solutions vanishes at one point in S, it vanishes everywhere in S. Moreover, if the Wronskian is non-zero at one point in S, it is non-zero everywhere in S. We also observe that $W(z)$ does not depend on the explicit functions $u_1(z)$ and $u_2(z)$, but only on $p(z)$. However, the functions $u_1(z)$ and $u_2(z)$ affect the constant $W(a)$.

Definition 2.3. Two functions, $u_1(z)$ and $u_2(z)$, are **linearly dependent** in a domain S (or an interval if we deal with functions defined on the real axis), if there exist constants c_1 and c_2, not both zero, such that

$$c_1 u_1(z) + c_2 u_2(z) = 0, \quad z \in S$$

If no such constants can be found, the functions are **linearly independent** in S.

The following lemma gives a check on linear independence.

Lemma 2.2. *Let $u_1(z)$ and $u_2(z)$ be two meromorphic functions in a domain S. A necessary and sufficient condition that $u_1(z)$ and $u_2(z)$ are linearly dependent is*

$$W(z) = 0, \quad z \in S$$

Proof. Without loss of generality, assume $u_1(z)$ not identically zero, and define $f(z) = u_2(z)/u_1(z)$. Then $f(z)$ is meromorphic in S, and $u_1(z)$ and $u_2(z)$ are linearly dependent if and only if $f(z)$ is constant. However, since $f(z)$ is meromorphic and

$$f'(z) = \frac{u_1(z)u_2'(z) - u_1'(z)u_2(z)}{(u_1(z))^2} = \frac{W(z)}{(u_1(z))^2}$$

$f(z)$ is constant if and only if $W(z) = 0$. $\square$

We apply the results to the differential equation (2.1) and assume we have two solutions, $u_1(z)$ and $u_2(z)$, to this differential equation in a domain S in the complex plane containing only regular points. If the functions $u_1(z)$ and $u_2(z)$ are linearly dependent, the Wronskian vanishes identically in the domain S, and if the Wronskian is non-zero at one point, the solutions $u_1(z)$ and $u_2(z)$ are independent in S, see Lemma 2.1.

2.3 Solution at a regular point

In a neighborhood of a regular point, a solution to the differential equation (2.1) can always be found. This section contains the details of such an explicit construction.

Let $z = b \in \mathbb{C}$ be a regular point, and let S_b be a circular neighborhood to b (radius $r_b > 0$) so that every point in S_b is a regular point, see Figure 2.1, i.e., take r_b smaller than the distance between b and the closest singular point P_1. We also assume $p(z) = 0$ — this is no restriction (transform to the reduced form).

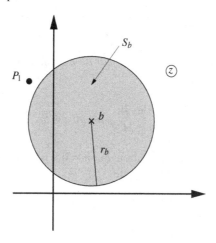

Fig. 2.1 The regular point $z = b$ and the circular domain S_b. The point P_1 denotes the closest singular point.

For given $a_0, a_1 \in \mathbb{C}$, define a sequence of functions $u_n(z)$, $n \in \mathbb{N}$, by the iteration scheme

$$\begin{cases} u_0(z) = a_0 + a_1(z - b), & a_0, a_1 \in \mathbb{C} \\ u_n(z) = \int_b^z (\zeta - z) q(\zeta) u_{n-1}(\zeta) \, d\zeta, & n \in \mathbb{Z}_+ \end{cases} \tag{2.3}$$

where the integration path is a straight line from b to z. We have the following fundamental theorem:

Theorem 2.2. *In a circular neighborhood, S_b, to the regular point $z = b \in \mathbb{C}$, the series*

$$u(z) = \sum_{n=0}^{\infty} u_n(z), \quad z \in S_b \tag{2.4}$$

obtained from the recursion formula in (2.3) is uniformly convergent and represents an analytic function in S_b. Specifically, $u(z)$ satisfies the differential equation

$$\frac{d^2 u(z)}{dz^2} + q(z) u(z) = 0 \tag{2.5}$$

with the following initial conditions at $z = b$:

$$\begin{cases} u(b) = a_0 \\ u'(b) = a_1 \end{cases} \tag{2.6}$$

Moreover, there is only one analytic function satisfying the differential equation (2.5) and the initial conditions (2.6), i.e., the function $u(z)$ in (2.4) is the unique analytic solution to the ODE, (2.5), with the given boundary conditions.

Proof. We prove this theorem in four lemmas, Lemmas 2.3–2.6. □

Lemma 2.3. *The sequence in* (2.3) *satisfies*

$$|u_n(z)| \le \mu M^n \frac{|z-b|^{2n}}{n!}, \quad \text{for all } z \in S_b, \text{ and } n \in \mathbb{Z}_+$$

where

$$\begin{cases} |u_0(z)| \le \mu \\ |q(z)| \le M \end{cases} \quad \text{for all } z \in S_b$$

Proof. We prove the lemma by induction. For $n = 0$ it is trivial. Let $n \ge 1$. We have, by the induction assumption, the estimate

$$|u_n(z)| \le \int_b^z |\zeta - z| M \mu M^{n-1} \frac{|\zeta - b|^{2n-2}}{(n-1)!} |d\zeta|, \quad z \in S_b$$

The integration path is: $\zeta(t) = (z-b)t + b, t \in [0,1]$. Therefore, since $|\zeta - z| \le |z-b|$ and $2n - 1 \ge n$, we get

$$|u_n(z)| \le \mu M^n |z-b| \int_0^1 \frac{|\zeta(t) - b|^{2n-2}}{(n-1)!} \left| \frac{d\zeta(t)}{dt} \right| dt$$

$$= \mu M^n |z-b|^2 \int_0^1 \frac{|z-b|^{2n-2} t^{2n-2}}{(n-1)!} dt = \mu M^n |z-b|^{2n} \int_0^1 \frac{t^{2n-2}}{(n-1)!} dt$$

$$= \mu M^n |z-b|^{2n} \frac{1}{(2n-1)(n-1)!} \le \mu M^n \frac{|z-b|^{2n}}{n!}$$

and the lemma is proved. □

Lemma 2.4. *The series*

$$u(z) = \sum_{n=0}^{\infty} u_n(z), \quad z \in S_b \tag{2.7}$$

is uniformly convergent in S_b, *and, therefore, analytic in* S_b.

Proof. From Lemma 2.3, we have

$$|u_n(z)| \le \mu M^n \frac{|z-b|^{2n}}{n!} \le \mu M^n \frac{r_b^{2n}}{n!}$$

Since the series

$$\sum_{n=0}^{\infty} \mu M^n \frac{r_b^{2n}}{n!} < \infty, \text{ for } r_b < \infty$$

the series (2.7) is uniformly convergent in S_b by the M-test of Weierstrass[5] [13, p. 107], and thus analytic in S_b. □

Lemma 2.5. *The analytic function* $u(z)$ *in Lemma 2.4 satisfies*

---

[5] Karl Weierstrass (1815–1897), German mathematician.

$$\frac{\mathrm{d}^2u(z)}{\mathrm{d}z^2}+q(z)u(z)=0$$

with the following initial conditions at $z = b$:

$$\begin{cases} u(b)=a_0 \\ u'(b)=a_1 \end{cases}$$

Proof. We easily get

$$\begin{cases} u'_n(z)=-\displaystyle\int_b^z q(\zeta)u_{n-1}(\zeta)\,\mathrm{d}\zeta \\ u''_n(z)=-q(z)u_{n-1}(z) \end{cases} \qquad n\in\mathbb{Z}_+$$

and, since the series (2.7) can be differentiated inside the sum, we get

$$\frac{\mathrm{d}^2u(z)}{\mathrm{d}z^2}=\underbrace{\frac{\mathrm{d}^2u_0(z)}{\mathrm{d}z^2}}_{=0}+\sum_{n=1}^{\infty}\frac{\mathrm{d}^2u_n(z)}{\mathrm{d}z^2}=-q(z)\sum_{n=1}^{\infty}u_{n-1}(z)=-q(z)u(z)$$

Moreover, the sequence takes at $z = b$ the values

$$\begin{cases} u_0(b)=a_0 \\ u_n(b)=0, \quad n\in\mathbb{Z}_+ \end{cases} \qquad \begin{cases} u'_0(b)=a_1 \\ u'_n(b)=0, \quad n\in\mathbb{Z}_+ \end{cases}$$

Therefore

$$\begin{cases} u(b)=\displaystyle\sum_{n=0}^{\infty}u_n(b)=a_0 \\ u'(b)=\displaystyle\sum_{n=0}^{\infty}u'_n(b)=a_1 \end{cases}$$

and the lemma is proved. □

Lemma 2.6. *There is only one analytic function $u(z)$ that satisfies*

$$\frac{\mathrm{d}^2u(z)}{\mathrm{d}z^2}+q(z)u(z)=0 \qquad \begin{cases} u(b)=a_0 \\ u'(b)=a_1 \end{cases}$$

Proof. Assume there are two analytic solutions, u_1 and u_2, and form $v = u_1 - u_2$. The function v is an analytic function and it satisfies

$$\frac{\mathrm{d}^2v(z)}{\mathrm{d}z^2}+q(z)v(z)=0 \qquad \begin{cases} v(b)=0 \\ v'(b)=0 \end{cases}$$

in S_b. By evaluating the differential equation at $z = b$, we obtain $v''(b) = 0$, since the point b is regular, i.e., $|q(b)| < \infty$. If we differentiate the equation, we get

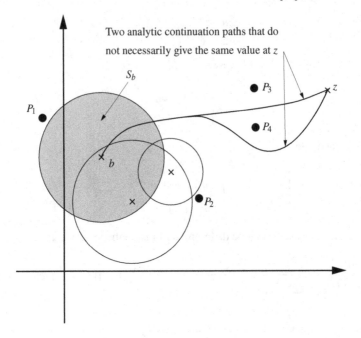

Fig. 2.2 The analytic continuation of the solution of the differential equation in the complex z-plane. The points P_i, $i = 1,2,3,4$, denote the singular points of the differential equation.

$$v'''(z) + q'(z)v(z) + q(z)v'(z) = 0$$

and evaluating this equation at $z = b$, we obtain by the same argument as above

$$v'''(b) = 0$$

Continuing the argument, we see that $v^{(n)}(b) = 0$ for all $n \in \mathbb{N}$. Since the solution $v(z)$ is analytic at $z = b$, we conclude that $v(z) = 0$ for all $z \in S_b$. $\square$

As a consequence of the Theorem 2.2, we can construct an analytic solution at any regular point to the ordinary differential equation in the complex z-plane. This means that we can construct a solution at all points in the complex z-plane, except at the singular points of the differential equation, where the solution shows singular behavior of some kind.

From a given initial condition, a unique solution is obtained by analytic continuation in the complex plane. The solution is unique at each regular point in the complex z-plane, but might depend on the way the analytic continuation is made, see Figure 2.2.

Any solution to (2.2), and therefore to (2.1), can be obtained as a linear combination of two linearly independent solutions to the equation. Two linearly independent solutions[6] $u_1(z)$ and $u_2(z)$ can be constructed by the initial conditions

$$\begin{cases} u_1(b) = a_0 = 0 \\ u_1'(b) = a_1 = 1 \end{cases} \qquad \begin{cases} u_2(b) = a_0 = 1 \\ u_2'(b) = a_1 = 0 \end{cases}$$

From the analysis in this section, we see that two linearly independent solutions can be constructed by analytic continuation everywhere in the complex plane, except at the singular points of the differential equation. The behavior near a singular point is, however, not determined by the analysis. In Section 2.4, we investigate the behavior of the solution near a singular point in more detail, but we finish this section with the construction of a second linearly independent solution if one solution is known.

A series solution of the solution at a regular point is investigated in Problem 2.6. This technique is exploited in more detail in Section 2.4.

2.3.1 The second solution

Now assume we have obtained one solution $u_1(z)$ to the differential equation — either by the explicit construction in Theorem 2.2 or by guessing or by some other way. In this section we explicitly construct a second linearly independent solution $u_2(z)$ from the solution $u_1(z)$. We proceed by writing the Wronskian in Definition 2.2 as

$$W(z) = u_1(z)u_2'(z) - u_1'(z)u_2(z) = (u_1(z))^2 \frac{\mathrm{d}}{\mathrm{d}z} \left(\frac{u_2(z)}{u_1(z)} \right)$$

Combine the equation with the result obtained in Lemma 2.1 into

$$\frac{\mathrm{d}}{\mathrm{d}z} \left(\frac{u_2(z)}{u_1(z)} \right) = \frac{W(a)}{(u_1(z))^2} \exp \left\{ - \int_a^z p(z'')\,\mathrm{d}z'' \right\}$$

and integrate in z from a to z. We get

$$\frac{u_2(z)}{u_1(z)} = W(a) \int_a^z \frac{1}{(u_1(z'))^2} \exp \left\{ - \int_a^{z'} p(z'')\,\mathrm{d}z'' \right\} \mathrm{d}z' + \frac{u_2(a)}{u_1(a)}$$

A second linearly independent solution to the differential equation then is

$$u_2(z) = u_1(z) \int^z \frac{1}{(u_1(z'))^2} \exp \left\{ - \int^{z'} p(z'')\,\mathrm{d}z'' \right\} \mathrm{d}z' \tag{2.8}$$

where we have dropped the lower limits (and the last constant term), which only give a term that is a proportional to the first solution $u_1(z)$, and, therefore, adds nothing

---

[6] The check of linear dependence is developed in Section 2.2.

new. In particular, for the reduced differential equation (2.2), where $p(z) = 0$, we have a particularly simple form of the second solution, viz.,

$$u_2(z) = u_1(z) \int^z \frac{dz'}{(u_1(z'))^2}$$

The method presented in this section has been employed to find a second solution to the differential equation in a neighborhood of a regular point, but the method may also have potential finding a second solution in the neighborhood of a singular point.

Example 2.1. We illustrate the result above with a very simple example. The differential equation

$$u''(z) - k^2 u(z) = 0$$

has a solution $u_1(z) = e^{kz}$. The second solution then is, $k \neq 0$

$$u_2(z) = u_1(z) \int^z \frac{dz'}{(u_1(z'))^2} = e^{kz} \int^z e^{-2kz'} dz' = -\frac{1}{2k} e^{kz} e^{-2kz} = -\frac{1}{2k} e^{-kz}$$

which, of course, is a second solution to the differential equation, which is linearly independent from the first one. ∎

Comment 2.1. The analysis in this section proved the existence of two linearly independent solutions $u_1(z)$ and $u_2(z)$ in the neighborhood of a regular point $z = b$. In Section 2.4, an explicit algorithm to find the power series solution is presented, in particular, see Comment 2.2 on page 23. ∎

2.4 Solution at a regular singular point

We have seen that, in general, two linearly independent solutions to the second order differential equation can be constructed at all points in $\mathbb{C}$ except at the singular points of the differential equation. In this section, we investigate the behavior of the solution near these singular points in detail. We start with a definition.[7]

Definition 2.4. Assume $z = c \in \mathbb{C}$ ($|c| < \infty$) is a singular point of the differential equation (2.1). The point $z = c$ is called a **regular singular point** of the differential equation (2.1) if $p(z)$ and $q(z)$ have the form

$$p(z) = \frac{P(z)}{z - c}, \qquad q(z) = \frac{Q(z)}{(z - c)^2}$$

where $P(z)$ and $Q(z)$ are analytic functions in a neighborhood of $z = c$. In all other cases, the singular point $z = c$ is called an **irregular singular point** of the differential equation (2.1).

---

[7] We use uppercase letters to denote functions that are analytic functions, and lowercase letters are used to denote meromorphic functions.

The point at infinity is special, and the classification of this point as a regular or an irregular singular point is postponed to Section 2.5.

2.4.1 The indicial equation

The appropriate differential equation for a regular singular point at $z = c$ therefore is

$$\frac{d^2 u(z)}{dz^2} + \frac{P(z)}{z-c}\frac{du(z)}{dz} + \frac{Q(z)}{(z-c)^2}u(z) = 0 \tag{2.9}$$

Since the functions $P(z)$ and $Q(z)$ are analytic in a neighborhood of $z = c$, i.e., they have power series expansions

$$\begin{cases} P(z) = \sum_{n=0}^{\infty} p_n(z-c)^n = p_0 + p_1(z-c) + p_2(z-c)^2 + \dots \\ Q(z) = \sum_{n=0}^{\infty} q_n(z-c)^n = q_0 + q_1(z-c) + q_2(z-c)^2 + \dots \end{cases} \tag{2.10}$$

which are convergent for all $z \in S_c$, where the set S_c is an open circle, radius $r_c > 0$, centered at $z = c$. The set S_c contains no other singular points of (2.9) than $z = c$.

We know that the solution $u(z)$ shows a singular behavior near the regular singular point $z = c \in \mathbb{C}$. To investigate this behavior, we make an ansatz for the solution[8]

$$\begin{aligned} u(z) &= (z-c)^\alpha \left(1 + \sum_{n=1}^{\infty} a_n(z-c)^n \right) \\ &= (z-c)^\alpha + a_1(z-c)^{\alpha+1} + a_2(z-c)^{\alpha+2} + \dots \end{aligned} \tag{2.11}$$

where, due to linearity and homogeneity of the differential equation, the first coefficient is set to 1, i.e., $a_0 = 1$. Our goal is to determine the unknown complex coefficients a_n, $n \in \mathbb{Z}_+$, provided the complex constants p_n and q_n, $n \in \mathbb{N}$, are known.

The differential equation (2.9) is employed, i.e.,

$$(z-c)^2 u''(z) + (z-c)P(z)u'(z) + Q(z)u(z) = 0$$

and we insert the solution $u(z)$. We get

$$(z-c)^\alpha \left\{ \alpha(\alpha-1) + \sum_{n=1}^{\infty} a_n(\alpha+n)(\alpha+n-1)(z-c)^n \right.$$
$$\left. + P(z)\left(\alpha + \sum_{n=1}^{\infty} a_n(\alpha+n)(z-c)^n \right) + Q(z)\left(1 + \sum_{n=1}^{\infty} a_n(z-c)^n \right) \right\} = 0$$

---

[8] This is the method of Frobenius named after the German mathematician Ferdinand Georg Frobenius (1849–1917).

Introduce the power series expansions of $P(z)$ and $Q(z)$ in (2.10), and identify the coefficient in front of each power of $z - c$ within the braces. Each of these coefficients must be zero if u is a solution to (2.9).

The result for each coefficient is

0) $\quad \alpha^2 + (p_0 - 1)\alpha + q_0 = 0$

1) $\quad a_1 \{(\alpha + 1)^2 + (p_0 - 1)(\alpha + 1) + q_0\} + p_1 \alpha + q_1 = 0$

2) $\quad a_2 \{(\alpha + 2)^2 + (p_0 - 1)(\alpha + 2) + q_0\} + a_1 \{p_1(\alpha + 1) + q_1\} + p_2 \alpha + q_2 = 0$

...

n) $\quad a_n \{(\alpha + n)^2 + (p_0 - 1)(\alpha + n) + q_0\}$

$$+ \sum_{m=1}^{n-1} a_{n-m} \{p_m(\alpha + n - m) + q_m\} + p_n \alpha + q_n = 0$$

We define the indicial equation

$$I(\alpha) = \alpha^2 + (p_0 - 1)\alpha + q_0 = 0 \tag{2.12}$$

and write the relations above as ($a_0 = 1$)

$$\begin{cases} I(\alpha) = 0 \\ a_1 I(\alpha + 1) + p_1 \alpha + q_1 = 0 \\ a_2 I(\alpha + 2) + a_1 \{p_1(\alpha + 1) + q_1\} + p_2 \alpha + q_2 = 0 \\ \cdots \\ a_n I(\alpha + n) + \sum_{m=1}^{n} a_{n-m} \{p_m(\alpha + n - m) + q_m\} = 0 \end{cases} \tag{2.13}$$

The coefficient in front of the lowest power is

$$I(\alpha) = 0$$

This indicial equation is a quadratic equation in α, and we denote the two roots of the equation by α_1 and α_2, respectively. Notice that the two roots satisfy

$$\alpha_1 + \alpha_2 = 1 - p_0, \qquad \alpha_1 \alpha_2 = q_0 \tag{2.14}$$

so that the values of $P(z)$ and $Q(z)$ at $z = c$ determine the roots of the indicial equation.

The coefficient in front of the second lowest power is

$$a_1 I(\alpha + 1) + p_1 \alpha + q_1 = 0$$

and, provided $I(\alpha + 1) \neq 0$, the coefficient a_1 has a unique solution in terms of the expansion coefficients of $P(z)$ and $Q(z)$. If we continue this argument, the coefficients a_n, $n \in \mathbb{Z}_+$, are uniquely soluble in terms of the expansion coefficients of

$P(z)$ and $Q(z)$, provided $I(\alpha + i) \neq 0$, $i = 1, 2, \ldots, n$. If this condition is met, the formal series in (2.11) can be constructed by the iterative scheme obtained from (2.13). We have

$$\begin{cases} a_0 = 1 \\ a_n = -\dfrac{\sum_{m=1}^{n} a_{n-m} \{p_m(\alpha + n - m) + q_m\}}{I(\alpha + n)}, \quad n = 1, 2, 3, \ldots \end{cases} \tag{2.15}$$

The following lemma quantifies when $I(\alpha + n)$ vanishes:

Lemma 2.7. *Denote by α_1 and α_2 the two roots of the indicial equation (2.12). Then*

$$I(\alpha_1 + n) = n(\alpha_1 - \alpha_2 + n)$$

Proof. Let $s = \alpha_1 - \alpha_2$. The lemma is then easily proved by the following observations:

$$\begin{aligned} I(\alpha_1 + n) &= (s + \alpha_2 + n)^2 + (p_0 - 1)(s + \alpha_2 + n) + q_0 \\ &= \underbrace{\alpha_2^2 + (p_0 - 1)\alpha_2 + q_0}_{=0} + (s + n)^2 + 2(s + n)\alpha_2 + (p_0 - 1)(s + n) \\ &= (s + n)(s + n + 2\alpha_2 + (p_0 - 1)) = n(s + n) = n(\alpha_1 - \alpha_2 + n) \end{aligned}$$

due to (2.14), and the lemma is proved. $\square$

From Lemma 2.7, we see that if the two roots of the indicial equation (2.12), α_1 and α_2, do not differ by an integer, then $I(\alpha_1 + n) \neq 0$, $n \in \mathbb{Z}_+$. The main investigation of the convergence of this power series, with coefficients obtained in (2.15), is postponed to Section 2.4.2. Meanwhile, we conclude that either of the following two cases can occur:

Case 1. $\alpha_1 - \alpha_2$ is not an integer. Then the iterative procedure in (2.15) has the potential of constructing two linearly independent solutions corresponding to the two roots $\alpha = \alpha_1$ and $\alpha = \alpha_2$, respectively.

Case 2. $\alpha_1 - \alpha_2$ is an integer. Then it is unclear whether the procedure gives a solution or not.

The growth rate of the coefficients a_n obtained from (2.15) is estimated in the next two lemmas.

Lemma 2.8. *Denote by α_1 and α_2 the two roots of the indicial equation (2.12), and let α_1 be the root with the largest real part, i.e., $\operatorname{Re} \alpha_1 \geq \operatorname{Re} \alpha_2$. Furthermore, assume*

$$|p_n| \leq \frac{M}{r^n}, \quad |p_n \alpha_1 + q_n| \leq \frac{M}{r^n}, \quad n \in \mathbb{Z}_+ \tag{2.16}$$

where $M > 1$. Then with $\alpha = \alpha_1$ in (2.15), the coefficients a_n satisfy

$$|a_n| \leq \frac{M^n}{r^n}, \quad n \in \mathbb{Z}_+$$

Proof. We prove the lemma by induction, and use Lemma 2.7, which with the notion $s = \alpha_1 - \alpha_2$ reads

$$I(\alpha_1 + n) = n(s + n) \tag{2.17}$$

The statement is true for $n = 1$, since, taking $\alpha = \alpha_1$ in (2.15), we get

$$|a_1| = \left| \frac{p_1 \alpha_1 + q_1}{I(\alpha_1 + 1)} \right| = \frac{|p_1 \alpha_1 + q_1|}{|s + 1|} \leq \frac{M}{r|s + 1|} \leq \frac{M}{r}$$

due to (2.16), (2.17), and $|s + 1| \geq 1$ (remember $\mathrm{Re}\, s \geq 0$, due to the assumption in the lemma).

Assume the induction statement is true for $k = 1, 2, \ldots, n - 1$. Then with $\alpha = \alpha_1$ in (2.15), we get for $n \geq 2$

$$
\begin{aligned}
|a_n| &= \left| \frac{\sum_{m=1}^{n} a_{n-m} \{ p_m(\alpha_1 + n - m) + q_m \}}{I(\alpha_1 + n)} \right| \\
&\leq \frac{\sum_{m=1}^{n} |a_{n-m}| \, |p_m \alpha_1 + q_m| + \sum_{m=1}^{n} |a_{n-m}| \, |p_m| (n - m)}{n|s + n|} \\
&\leq \frac{\sum_{m=1}^{n} M^{n-m} r^{m-n} M r^{-m} + \sum_{m=1}^{n} M^{n-m} r^{m-n} M r^{-m} (n - m)}{n^2 |1 + s/n|}
\end{aligned}
$$

by (2.16), (2.17), and the induction assumption. Further estimates give (remember $M > 1$)

$$|a_n| \leq \frac{M^n}{r^n} \frac{n + \sum_{m=1}^{n}(n - m)}{n^2 |1 + s/n|} = \frac{M^n}{r^n} \frac{n + n(n-1)/2}{n^2 |1 + s/n|} = \frac{M^n}{r^n} \frac{n + 1}{2n|1 + s/n|}$$

and since $|1 + s/n| \geq 1$ (remember $\mathrm{Re}\, s \geq 0$), and $n + 1 \leq 2n$, we have

$$|a_n| \leq \frac{M^n}{r^n}, \quad n \geq 2$$

and the induction proof is finished. $\square$

The growth rate of the coefficients a_n corresponding to the other root is more complex.

Lemma 2.9. *Denote by α_1 and α_2 the two roots of the indicial equation (2.12), and let α_2 be the root with the smallest real part, i.e., $\mathrm{Re}\, \alpha_1 \geq \mathrm{Re}\, \alpha_2$. Moreover, assume $s = \alpha_1 - \alpha_2 \notin \mathbb{Z}_+$, and*

$$|p_n| \leq \frac{M}{r^n}, \quad |p_n \alpha_2 + q_n| \leq \frac{M}{r^n}, \quad n \in \mathbb{Z}_+ \tag{2.18}$$

Then with $\alpha = \alpha_2$ in (2.15), the coefficients a_n satisfy

$$|a_n| \leq \frac{M'^n}{r^n}, \quad n \in \mathbb{Z}_+$$

where $M' = M/\kappa$ and $M' > 1$. Here $\kappa = \inf\{|1 - s|, |1 - s/2|, |1 - s/3|, \ldots\}$ is a positive number, due to the assumptions on s.

Proof. Again, we prove by induction, and use Lemma 2.7, which with the notion $s = \alpha_1 - \alpha_2$ reads

$$I(\alpha_2 + n) = n(n - s) \tag{2.19}$$

We conclude that the statement is true for $n = 1$, since, taking $\alpha = \alpha_2$ in (2.15), we get

$$|a_1| = \left|\frac{p_1\alpha_2 + q_1}{I(\alpha_2 + 1)}\right| = \frac{|p_1\alpha_2 + q_1|}{|1 - s|} \leq \frac{M}{r|1 - s|} \leq \frac{M}{\kappa r} = \frac{M'}{r}$$

due to (2.18), (2.19), and $|1 - s| \geq \kappa$.

Assume the induction statement is true for $k = 1, 2, \ldots, n - 1$. Then with $\alpha = \alpha_2$ in (2.15), we get for $n \geq 2$

$$|a_n| = \left|\frac{\sum_{m=1}^{n} a_{n-m}\{p_m(\alpha_2 + n - m) + q_m\}}{I(\alpha_2 + n)}\right|$$

$$\leq \frac{\sum_{m=1}^{n} |a_{n-m}| |p_m\alpha_2 + q_m| + \sum_{m=1}^{n} |a_{n-m}| |p_m| (n - m)}{n|n - s|}$$

$$\leq \frac{\sum_{m=1}^{n} M'^{n-m} r^{m-n} M r^{-m} + \sum_{m=1}^{n} M'^{n-m} r^{m-n} M r^{-m} (n - m)}{n^2 |1 - s/n|}$$

by (2.18), (2.19), and the induction assumption. Further estimates give (remember $M' > 1$)

$$|a_n| \leq \frac{M M'^{n-1}}{r^n} \frac{n + \sum_{m=1}^{n} (n - m)}{n^2 |1 - s/n|}$$

$$= \frac{M M'^{n-1}}{r^n} \frac{n + n(n - 1)/2}{n^2 |1 - s/n|} = \frac{M M'^{n-1}}{r^n} \frac{n + 1}{2n|1 - s/n|}$$

and since $|1 - s/n| \geq \kappa$ and $n + 1 \leq 2n$, we have

$$|a_n| \leq \frac{M^n}{\kappa^n r^n} \leq \frac{M'^n}{r^n}$$

and the induction proof is finished. □

2.4.2 Convergence of the solution

We now address the question of convergence of the power series

$$\sum_{n=1}^{\infty} a_n (z - c)^n$$

with the coefficients a_n, $n \in \mathbb{Z}_+$, explicitly constructed by the algorithm in Section 2.4.1 above. The main theorem of this section is:

Theorem 2.3. *Denote by α_1 and α_2 the two roots of the indicial equation (2.12), and number the roots such that $\operatorname{Re}\alpha_1 \geq \operatorname{Re}\alpha_2$. Then the power series*

$$\sum_{n=1}^{\infty} a_n(z-c)^n$$

with coefficients, a_n, obtained by the iteration scheme in (2.15), starting from the root α_1, represents an analytic function in a neighborhood of the regular singular point $z = c$.

Moreover, if the two roots α_1 and α_2 of the indicial equation do not differ by a positive integer,[9] the power series

$$\sum_{n=1}^{\infty} a'_n(z-c)^n$$

with coefficients, a'_n, obtained by the iteration scheme in (2.15), starting from the root α_2, also represents an analytic function in a neighborhood of the regular singular point $z = c$.

The theorem guarantees that there exists at least one solution to the differential equation, (2.9), in a neighborhood of the regular singular point, $z = c$. If the the roots of the indicial equations do not differ by a positive integer, we can also construct a second solution. In this case, the solutions are

$$u_1(z) = (z-c)^{\alpha_1}\left(1 + \sum_{n=1}^{\infty} a_n(z-c)^n\right), \quad u_2(z) = (z-c)^{\alpha_2}\left(1 + \sum_{n=1}^{\infty} a'_n(z-c)^n\right)$$

Proof. The explicit coefficients a_n, obtained by the algorithm in Section 2.4.1, is given by (2.15), i.e., $I(\alpha) = 0$ and

$$\begin{cases} a_0 = 1 \\ a_n = -\dfrac{\sum_{m=1}^{n} a_{n-m}\{p_m(\alpha+n-m)+q_m\}}{I(\alpha+n)}, & n = 1,2,3,\dots \end{cases} \tag{2.20}$$

To analyze the convergence, denote the two roots of the indicial equation by α_1 and α_2, respectively, and let $s = \alpha_1 - \alpha_2$. Number the roots such that $\operatorname{Re}s \geq 0$, i.e., $\operatorname{Re}\alpha_1 \geq \operatorname{Re}\alpha_2$. The assumption of the second part of the theorem implies that $s \notin \mathbb{Z}_+$.

The coefficients in the power series expansions of the functions $P(z)$ and $Q(z)$ in (2.10), i.e.,

---

[9] If both roots are identical, $\alpha_1 = \alpha_2$, only one solution is obtained.

$$
\begin{cases}
P(z) = \displaystyle\sum_{n=0}^{\infty} p_n(z-c)^n \\[4mm]
Q(z) = \displaystyle\sum_{n=0}^{\infty} q_n(z-c)^n
\end{cases}
$$

are determined by[10]

$$
p_n = \frac{1}{2\pi i} \oint_C \frac{P(t)\,dt}{(t-c)^{n+1}}, \qquad q_n = \frac{1}{2\pi i} \oint_C \frac{Q(t)\,dt}{(t-c)^{n+1}}
$$

where C is a circle with radius r_c, centered at $z = c$, and contained in the common domain of analyticity of $P(z)$ and $Q(z)$. Let the constant M_p and M_q be the maximum value of $|P(z)|$ and $|Q(z)|$, respectively, on the circle C. Then

$$
|p_n| \le \frac{M_p}{r_c^n}, \qquad |q_n| \le \frac{M_q}{r_c^n}, \qquad n \in \mathbb{Z}_+
$$

The coefficients p_0 and q_0 satisfy similar inequalities, but these are not used in the proof.

Denote by $M = \max\{M_p, |\alpha_1|M_p + M_q, |\alpha_2|M_p + M_q\}$, which is a finite number, due to the assumptions made on $P(z)$ and $Q(z)$. Then, we have for $n \in \mathbb{Z}_+$

$$
|p_n| \le \frac{M}{r_c^n}, \qquad |p_n\alpha_1 + q_n| \le \frac{M}{r_c^n}, \qquad |p_n\alpha_2 + q_n| \le \frac{M}{r_c^n}, \tag{2.21}
$$

There is no loss of generality to assume that the constant $M > 1$ — increase the value of M if necessary.

From Lemma 2.8, we then obtain

$$
|a_n| \le \frac{M^n}{r_c^n}, \qquad n \in \mathbb{Z}_+
$$

if we are using the root α_1 — the one with the largest real part. The power series

$$
\sum_{n=1}^{\infty} a_n(z-c)^n
$$

is then uniformly convergent inside the circle $|z-c| < \rho = r_c/M$ since

---

[10] The expansion coefficients of an analytic function $f(z)$ in a power series expansion satisfy [13, p. 128]

$$
f(z) = \sum_{n=0}^{\infty} \frac{f^{(n)}(c)}{n!}(z-c)^n
$$

where [13, p. 180]

$$
f^{(n)}(c) = \frac{n!}{2\pi i} \oint_C \frac{f(t)\,dt}{(t-c)^{n+1}}, \qquad n \in \mathbb{Z}_+
$$

$$\sum_{n=1}^{\infty} |a_n||z-c|^n \leq \sum_{n=1}^{\infty} \frac{M^n|z-c|^n}{r_c^n} = \sum_{n=1}^{\infty} \frac{|z-c|^n}{\rho^n} < \infty$$

and the series represents an analytic function inside the circle $|z-c| < \rho$, and the first part of the theorem is proved.

Using the second root of the indicial equation, α_2, and Lemma 2.9, we obtain

$$|a_n'| \leq \frac{M'^n}{r_c^n} = \frac{M^n}{\kappa^n r_c^n}, \quad n \in \mathbb{Z}_+$$

where $\kappa = \inf\{|1-s|, |1-s/2|, |1-s/3|, \ldots\}$, which is a positive number, due to the assumption that s is not a positive integer, and $M' = M/\kappa$. The power series

$$\sum_{n=1}^{\infty} a_n'(z-c)^n$$

is then uniformly convergent inside the circle $|z-c| < \rho' = r_c/M' = \kappa r_c/M$ since

$$\sum_{n=1}^{\infty} |a_n'||z-c|^n \leq \sum_{n=1}^{\infty} \frac{M^n|z-c|^n}{\kappa^n r_c^n} = \sum_{n=1}^{\infty} \frac{|z-c|^n}{\rho'^n} < \infty$$

and the series represents an analytic function inside the circle $|z-c| < \rho'$. This completes the second part of the proof of the theorem. $\square$

Indeed, if $\alpha_1 - \alpha_2$ is not a positive integer or zero, we have constructed two linearly independent[11] solutions of the differential equation, viz.,

$$\begin{cases} u_1(z) = (z-c)^{\alpha_1} \left(1 + \sum_{n=1}^{\infty} a_n(z-c)^n \right) \\ u_2(z) = (z-c)^{\alpha_2} \left(1 + \sum_{n=1}^{\infty} a_n'(z-c)^n \right) \end{cases} \tag{2.22}$$

2.4.3 The second solution — exceptional case

The existence of one solution to the differential equation of the form, see (2.22),

$$u_1(z) = (z-c)^{\alpha_1} \left(1 + \sum_{n=1}^{\infty} a_n(z-c)^n \right) = (z-c)^{\alpha_1} f(z) \tag{2.23}$$

where $f(z)$ is analytic in a neighborhood of the regular singular point $z = c$, is guaranteed by Theorem 2.3. It is constructed from the root of the indicial equation with the largest real part.

---

[11] They are linearly independent since they have different analytic properties at $z = c$.

In this section, we construct a second, linearly independent solution, $u_2(z)$, to the differential equation, in a neighborhood of the regular singular point $z = c$, when $s = \alpha_1 - \alpha_2 \in \mathbb{N}$, i.e., it is a positive integer or zero. This is exactly the case when the construction of the power series of second solution, $u_2(z)$, in the proof of Theorem 2.3 breaks down or only gives one solution. The reason for the failure is that the quantity κ in the proof then is zero, and the constructions fail due to division by zero. The first solution $u_1(z)$ in (2.23) is used to construct the second solution.

In Section 2.3.1, we presented a method to find a second solution, $u_2(z)$, if one solution, $u_1(z)$, is known. Here, we prefer to employ a variation of this method, which is more pertinent for the analysis in this section. The method is also called "variation of the constant," and we proceed, by making a formal ansatz

$$u_2(z) = C(z)u_1(z)$$

Our aim is to find the conditions on $C(z)$ that make $u_2(z)$ a solution to the differential equation (2.9), if $u_1(z)$ is a known solution of the same differential equation.

Differentiate the solution $u_2(z)$ and insert in the differential equation (2.9). This implies

$$C''(z)u_1(z) + 2C'(z)u_1'(z) + C(z)u_1''(z)$$
$$+ \frac{P(z)}{z-c}C'(z)u_1(z) + \frac{P(z)}{z-c}C(z)u_1'(z) + \frac{Q(z)}{(z-c)^2}C(z)u_1(z) = 0$$

If $u_1(z)$ is the solution given in (2.23), then it satisfies (2.9), and the expression above simplifies to

$$C''(z)u_1(z) + 2C'(z)u_1'(z) + \frac{P(z)}{z-c}C'(z)u_1(z) = 0$$

or

$$\frac{C''(z)}{C'(z)} + 2\frac{u_1'(z)}{u_1(z)} + \frac{P(z)}{z-c} = 0$$

We easily integrate this expression and get

$$\ln C'(z) + 2\ln u_1(z) + \int^z \frac{P(z')}{z'-c}\,dz' = A$$

where A is a complex constant. The integral becomes

$$\int^z \frac{P(z')}{z'-c}\,dz' = \int^z \frac{p_0 + p_1(z'-c) + p_2(z'-c)^2 + \cdots}{z'-c}\,dz' = p_0\ln(z-c) + F(z)$$

where $F(z)$ is analytic[12] in a neighborhood of $z = c$. The expression above then is

---

[12] Note that

$$\int^z \sum_{i=0}^{\infty} p_{1+i}(z-c)^i\,dz' = \sum_{i=0}^{\infty} \frac{p_{1+i}}{i+1}(z-c)^{i+1} + C$$

$$\ln C'(z) + 2\ln u_1(z) + \ln(z-c)^{p_0} + F(z) = A$$

or[13]

$$C(z) = \int^z \frac{G(z')}{(u_1(z'))^2 (z'-c)^{p_0}} \, dz' = \int^z \frac{H(z')}{(z'-c)^{2\alpha_1 + p_0}} \, dz'$$

where $G(z) = \exp\{A - F(z)\}$ and $H(z) = G(z)/(f(z))^2$ are analytic functions in a neighborhood of $z = c$ (remember $f(z)$ is non-zero in a neighborhood of $z = c$).

The roots of the indicial equations satisfy $\alpha_1 + \alpha_2 = 1 - p_0$, see (2.14). The function $C(z)$ then is ($s = \alpha_1 - \alpha_2$ is a positive integer or zero)

$$C(z) = \int^z \frac{H(z')}{(z'-c)^{s+1}} \, dz' = \int^z \frac{h_0 + h_1(z'-c) + h_2(z'-c)^2 + \cdots}{(z'-c)^{s+1}} \, dz'$$

where the power series expansion of $H(z)$

$$H(z) = \sum_{n=0}^{\infty} h_n (z-c)^n$$

converges in a neighborhood of $z = c$. Two different cases appear.

Case 1. First, if $s = 0$, i.e., the two roots of the indicial equations coincide, then

$$C(z) = \int^z \frac{h_0 + h_1(z'-c) + h_2(z'-c)^2 + \cdots}{z'-c} \, dz' = h_0 \ln(z-c) + K_1(z)$$

where $K_1(z)$ is analytic[14] in a neighborhood of $z = c$. The second solution $u_2(z)$ can then be written as

$$u_2(z) = C(z)u_1(z) = (h_0 \ln(z-c) + K_1(z))(z-c)^{\alpha_1} f(z)$$

where $f(z)$ and $K_1(z)$ are analytic functions in a neighborhood of $z = c$.

Case 2. The second case appears when s is a positive integer, i.e., $s = n \in \mathbb{Z}_+$. Then we get

$$C(z) = \int^z \frac{h_0 + h_1(z'-c) + h_2(z'-c)^2 + \cdots}{(z'-c)^{n+1}} \, dz'$$

$$= \int^z \left(\sum_{i=0}^{n-1} \frac{h_i}{(z-c)^{n+1-i}} + \frac{h_n}{z-c} + \sum_{i=0}^{\infty} h_{n+1+i}(z-c)^i \right) dz'$$

which we simplify to

---

has the same radius of convergence as the power series of $P(z)$.

[13] Notice the resemblance with the result in (2.8) in Section 2.3.1.

[14] The change of order between summation and integration, and the radius of convergence of the power series of $H(z)$, see footnote 12 above.

$$C(z) = -\sum_{i=0}^{n-1} \frac{h_i(z-c)^{i-n}}{n-i} + h_n \ln(z-c) + K_2(z)$$

$$= (z-c)^{-n} K_3(z) + h_n \ln(z-c)$$

where $K_2(z)$ and $K_2(z)$ are analytic in a neighborhood of $z = c$. The second solution $u_2(z)$ can then be written as

$$u_2(z) = C(z)u_1(z) = \left((z-c)^{-n} K_3(z) + h_n \ln(z-c)\right)(z-c)^{\alpha_1} f(z)$$

where $f(z)$ and $K_3(z)$ are analytic functions in a neighborhood of $z = c$.

We conclude that, in both cases ($s = 0$ and $s = n \in \mathbb{Z}_+$), a logarithmic term appears, and the second solution can be written (remember $s = \alpha_1 - \alpha_2$)

$$u_2(z) = (z-c)^{\alpha_2} g(z) + h_n \ln(z-c)(z-c)^{\alpha_1} f(z)$$

where

$$f(z) = 1 + \sum_{n=1}^{\infty} a_n(z-c)^n$$

and the coefficients a_n are determined above in Theorem 2.3, and $g(z)$ is an analytic function in a neighborhood of $z = c$.

For convenience, we summarize the results in this section in a theorem.

Theorem 2.4. *Denote by α_1 and α_2 the two roots of the indicial equation (2.12) to the differential equation (2.9), and number the roots such that $\mathrm{Re}\,\alpha_1 \geq \mathrm{Re}\,\alpha_2$. Then the two linearly independent solutions to (2.9) are:*

1. If $s = \alpha_1 - \alpha_2 \notin \mathbb{N}$

$$\begin{cases} u_1(z) = (z-c)^{\alpha_1} f_1(z) \\ u_2(z) = (z-c)^{\alpha_2} f_2(z) \end{cases}$$

where $f_1(z)$ and $f_2(z)$ are analytic functions in a neighborhood of $z = c$. The explicit constructions of these analytic functions are made in Theorem 2.3.
2. If $s = \alpha_1 - \alpha_2 \in \mathbb{N}$

$$\begin{cases} u_1(z) = (z-c)^{\alpha_1} f_1(z) \\ u_2(z) = (z-c)^{\alpha_2} g(z) + h_n \ln(z-c)(z-c)^{\alpha_1} f_1(z) \end{cases}$$

where $f_1(z)$ and $g(z)$ are analytic functions in a neighborhood of $z = c$. The explicit construction of $f_1(z)$ is made in Theorem 2.3.

Comment 2.2. The construction of the solutions $u_1(z)$ and $u_2(z)$ in a neighborhood of a regular singular point $z = c$, see Theorem 2.4, can, of course, also be applied if the point $z = c$ is regular. This becomes a special case of the above, simply by letting $p_0 = q_0 = q_1 = 0$. We then apply the result of the theorem with the roots of the indicial equation $\alpha_1 = 1$ and $\alpha_2 = 0$. The iterative scheme in (2.15) corresponding to $\alpha_1 = 1$ then becomes (Lemma 2.7 implies $I(\alpha_1 + n) = n(n+1)$)

$$
\begin{cases}
a_1 = -\dfrac{p_1}{2} \\[2mm]
a_2 = -\dfrac{a_1 p_1 + p_2 + q_2}{6} = \dfrac{p_1^2/2 - p_2 - q_2}{6} \\[2mm]
a_n = -\dfrac{\sum_{m=1}^{n} a_{n-m}\{p_m(1+n-m) + q_m\}}{n(n+1)}, \quad n = 2, 3, 4, \ldots
\end{cases}
$$

Due to Theorem 2.4, the power series solution $u_1(z)$ given by

$$
u_1(z) = (z - c)\left(1 + \sum_{n=1}^{\infty} a_n(z - c)^n\right)
$$

is convergent in a neighborhood of the regular point $z = c$. Also, compare this result with the result of Problem 2.6. ∎

2.5 Solution at a regular singular point at infinity

The point at infinity is different from all points in the finite complex plane, and it has to be analyzed as a special case. To find out the behavior of the differential equation at the point at infinity, introduce a new variable ζ, defined as

$$
\zeta = \frac{1}{z} \implies \frac{d}{dz} = -\zeta^2 \frac{d}{d\zeta} \quad \text{and} \quad \frac{d^2}{dz^2} = \zeta^2 \frac{d}{d\zeta} \zeta^2 \frac{d}{d\zeta} = \zeta^4 \frac{d^2}{d\zeta^2} + 2\zeta^3 \frac{d}{d\zeta}
$$

The point at $z = \infty$ is then mapped to $\zeta = 0$, and in this new variable, ζ, we can apply the results already obtained in the sections above to the origin $\zeta = 0$.

In fact, the differential equation (2.1) becomes

$$
\zeta^4 \frac{d^2 u}{d\zeta^2} + (2\zeta^3 - \zeta^2 p(1/\zeta)) \frac{du}{d\zeta} + q(1/\zeta)u = 0
$$

or

$$
\frac{d^2 u}{d\zeta^2} + \frac{1}{\zeta}\left(2 - \frac{1}{\zeta} p(1/\zeta)\right)\frac{du}{d\zeta} + \frac{1}{\zeta^4} q(1/\zeta)u = 0 \tag{2.24}
$$

and the behavior of the original equation at $z = \infty$ is transformed to an ordinary differential equation at a finite point $\zeta = 0$, which we know how to handle and classify.

We proceed by investigating the conditions that the coefficients have to satisfy in order to have a regular point, a regular singular point, or an irregular singular point at infinity, respectively. We state the result as a definition.

Definition 2.5. The point at infinity is classified as:

1. The point $z = \infty$ is a **regular point** of the differential equation (2.1) provided

$$\frac{2}{\zeta} - \frac{1}{\zeta^2} p(1/\zeta) \text{ is analytic at } \zeta = 0 \quad \Longleftrightarrow \quad 2z - z^2 p(z) \text{ is analytic at } z = \infty$$

which means that $p(z) = 2/z + O(z^{-2})$ as $z \to \infty$,

and

$$\frac{1}{\zeta^4} q(1/\zeta) \text{ is analytic at } \zeta = 0 \quad \Longleftrightarrow \quad z^4 q(z) \text{ is analytic at } z = \infty$$

which means that $q(z) = O(z^{-4})$ as $z \to \infty$.

2. The point $z = \infty$ is a **regular singular point** of the differential equation (2.1) provided

$$2 - \frac{1}{\zeta} p(1/\zeta) \text{ is analytic at } \zeta = 0 \quad \Longleftrightarrow \quad zp(z) \text{ is analytic at } z = \infty$$

which means that $p(z) = O(z^{-1})$ as $z \to \infty$,

and

$$\frac{1}{\zeta^2} q(1/\zeta) \text{ is analytic at } \zeta = 0 \quad \Longleftrightarrow \quad z^2 q(z) \text{ is analytic at } z = \infty$$

which means that $q(z) = O(z^{-2})$ as $z \to \infty$.

3. The point $z = \infty$ is an **irregular singular point** of the differential equation (2.1) in all other cases.

The analysis of a regular or singular point at infinity therefore is transformed to an investigation of the properties at the origin in the ζ variable.

Example 2.2. The differential equation

$$\frac{d^2 u(z)}{dz^2} + \frac{2}{z} \frac{du(z)}{dz} + \frac{1}{z^2} u(z) = 0$$

has a singular point at $z = \infty$. This point is a regular singular point. ∎

Example 2.3. The differential equation

$$\frac{d^2 u(z)}{dz^2} + \frac{2}{z} \frac{du(z)}{dz} + u(z) = 0$$

has a singular point at $z = \infty$. This point is an irregular singular point. ∎

Problems

2.1. Prove the result of Theorem 2.1.

2.2. Prove, using the result of Problem 2.1, that if $u_1(z)$ is a solution (2.2)

$$\frac{d^2 u_1(z)}{dz^2} + q_1(z)u_1(z) = 0$$

where $q_1(z)$ is analytic in a domain S, then (b is a regular point of the differential equation)

$$u(z) = u_1(z)\exp\left\{-\frac{1}{2}\int_b^z p(z')\,dz'\right\}$$

solves (2.1)

$$\frac{d^2 u(z)}{dz^2} + p(z)\frac{du(z)}{dz} + q(z)u(z) = 0$$

and vice versa, provided

$$q_1(z) = q(z) - \frac{1}{2}\frac{dp(z)}{dz} - \frac{1}{4}(p(z))^2$$

2.3. Check that

$$u_1(z) = \frac{\sin z}{z}$$

solves

$$\frac{d^2 u_1(z)}{dz^2} + \frac{2}{z}\frac{du_1(z)}{dz} + u_1(z) = 0$$

Find a second, linearly independent solution to the differential equation.

2.4. One form of Lamé's[15] differential equation reads ($a \neq b$ and $a, b \neq 0$)[16]

$$\frac{d^2 u(z)}{dz^2} + \left(\frac{z}{z^2 - a^2} + \frac{z}{z^2 - b^2}\right)\frac{du(z)}{dz} + \frac{k - m(m+1)z^2}{(z^2 - a^2)(z^2 - b^2)}u(z) = 0$$

Find its singular points and classify them, and determine the roots of the indicial equation. Moreover, find the Wronskian $W(z)$ of the two linearly independent solutions, $u_1(z)$ and $u_2(z)$, that satisfy[17]

$$\begin{cases} u_1(0) = 1 \\ u_1'(0) = 0 \end{cases} \quad \begin{cases} u_2(0) = 0 \\ u_2'(0) = 1 \end{cases}$$

---

[15] Gabriel Lamé (1795–1870), French mathematician.

[16] Lamé's differential equation is also briefly mentioned in Section 8.6.1.

[17] A way of constructing these solutions is to use the result of Problem 2.6.

2.5. Electromagnetic scattering by a radially inhomogeneous sphere leads to a differential equation of the form

$$\frac{d^2u(z)}{dz^2} + \left(q_1(z) - \frac{l(l+1)}{z^2} \right) u(z) = 0$$

where q_1 is an analytic function everywhere in the finite complex plane, and $q_1(z) \to$ constant as $|z| \to \infty$ and $l \in \mathbb{Z}_+$. Classify the points to the differential equation, and find the functional behavior of its solutions at the origin.

2.6. [†]Let $z = b$ be a regular point to the differential equation

$$\frac{d^2u(z)}{dz^2} + p(z)\frac{du(z)}{dz} + q(z)u(z) = 0$$

where $p(z)$ and $q(z)$ have power series expansions

$$\begin{cases} p(z) = \sum_{n=0}^{\infty} p_n(z-b)^n = p_0 + p_1(z-b) + p_2(z-b)^2 + \ldots \\ q(z) = \sum_{n=0}^{\infty} q_n(z-b)^n = q_0 + q_1(z-b) + q_2(z-b)^2 + \ldots \end{cases}$$

which are convergent in a neighborhood of $z = b$. Assume a power series expansion of the solution $u(z)$

$$u(z) = \sum_{n=0}^{\infty} a_n(z-b)^n = a_0 + a_1(z-b) + a_2(z-b)^2 + \ldots$$

and write the solution as

$$u(z) = a_0 u_1(z) + a_1 u_2(z)$$

Find the power series of $u_1(z)$ and $u_2(z)$, and prove that the series converge, and that the solutions $u_1(z)$ and $u_2(z)$ are linearly independent.
Hint: Copy relevant parts of Section 2.4, and use the result of Theorem 2.2.

Chapter 3
Equations of Fuchsian type

Again, we focus on the standard form of the ordinary differential equation of the second order, viz., equation (2.1)

$$\frac{d^2u(z)}{dz^2} + p(z)\frac{du(z)}{dz} + q(z)u(z) = 0$$

If all its singular points are regular singular points (including the point at infinity), see Definitions 2.4 and 2.5 on pages 12 and 24, respectively, the equation is of **Fuchsian type**.[1] An equation of Fuchsian type therefore only has regular and regular singular points in the complex plane (including the point at infinity). This implies — as we shall see soon — that the functions $p(z)$ and $q(z)$ are rational functions. We divide the analysis below into two different cases, Section 3.1 and Section 3.2, depending on whether the point at infinity is a singular regular point or a regular point, respectively. The chapter ends with the displacement theorem in Section 3.3.

3.1 Regular singular point at infinity

We start this section with an investigation of the form of the functions $p(z)$ and $q(z)$, and the results are then collected in a theorem on page 32.

We assume there are n distinct, regular singular points in the finite complex plane, located at $z = a_r$, $r = 1, 2, 3, \ldots, n$, and one at $z = \infty$. Moreover, no irregular singular points are present anywhere in the complex plane. The roots of the indicial equation are denoted by α_r, β_r, $r = 1, 2, 3, \ldots, n$, and at infinity we denote the roots by α, β. The functions $p(z)$ and $q(z)$ then must be of the form, see Definition 2.4 on page 12

$$p(z) = \frac{G(z)}{\psi(z)}, \qquad q(z) = \frac{H(z)}{(\psi(z))^2}$$

---

[1] Lazarus Fuchs (1833–1902), German mathematician.

G. Kristensson, *Second Order Differential Equations: Special Functions and Their Classification*, DOI 10.1007/978-1-4419-7020-6_3,
© Springer Science+Business Media, LLC 2010

where

$$\psi(z) = \prod_{r=1}^{n}(z - a_r),$$

and $G(z)$ and $H(z)$ are analytic functions everywhere in the complex plane (entire functions). Since the condition of a regular singular point at infinity implies that $zp(z)$ and $z^2q(z)$ are analytic at $z = \infty$, see Definition 2.5 on page 24, we conclude that $p(z)$ and $q(z)$ do not grow faster than $|z|^{-1}$ and $|z|^{-2}$, respectively, as $z \to \infty$. Recall that

$$G(z) = p(z) \prod_{r=1}^{n}(z - a_r) \quad \text{and} \quad H(z) = q(z) \prod_{r=1}^{n}(z - a_r)^2$$

are entire functions, and $G(z)$ and $H(z)$ do not grow faster than $|z|^{n-1}$ and $|z|^{2n-2}$, respectively, as $z \to \infty$. Therefore, they both are polynomials of degree $n - 1$ and $2n - 2$, respectively, due to Liouville's[2] theorem [9, p. 87] or [13, p. 204]. We have

$$\begin{cases} G(z) \text{ is a polynomial of degree } \leq n - 1 \\ H(z) \text{ is a polynomial of degree } \leq 2n - 2 \end{cases}$$

Using partial fraction decomposition,[3] we write the functions $p(z)$ and $q(z)$ as

$$p(z) = \sum_{r=1}^{n} \frac{c_r}{z - a_r} \tag{3.1}$$

and either of the two representations of $q(z)$

$$q(z) = \sum_{r=1}^{n} \frac{d_r}{(z - a_r)^2} + \frac{R(z)}{\psi(z)} = \sum_{r=1}^{n} \frac{d_r}{(z - a_r)^2} + \sum_{r=1}^{n} \frac{f_r}{z - a_r} \tag{3.2}$$

where $R(z)$ is a polynomial of degree $n - 2$, satisfying one condition, see below. The leading coefficients in these representations are determined by

$$\begin{cases} c_i = \lim_{z \to a_i}(z - a_i)p(z) \\ d_i = \lim_{z \to a_i}(z - a_i)^2 q(z) \end{cases} \quad i = 1, 2, 3, \ldots, n$$

Focus on the singularity at a specific singular point, say $z = a_i$, and write $p(z)$ and $q(z)$ as

$$\begin{cases} p(z) = \dfrac{P(z)}{z - a_i} = \dfrac{1}{z - a_i} \sum_{n=0}^{\infty} p_n(z - a_i)^n \\ q(z) = \dfrac{Q(z)}{(z - a_i)^2} = \dfrac{1}{(z - a_i)^2} \sum_{n=0}^{\infty} q_n(z - a_i)^n \end{cases}$$

---

[2] Joseph Liouville (1809–1882), French mathematician.
[3] Some results on partial fractions are presented in Appendix C, see also [10, p. 217–218]

The roots of the indicial equation at $z = a_i$ satisfy (2.14)

$$\begin{cases} \alpha_i + \beta_i = 1 - p_0 = 1 - \lim_{z \to a_i} (z - a_i) p(z) = 1 - c_i \\ \alpha_i \beta_i = q_0 = \lim_{z \to a_i} (z - a_i)^2 q(z) = d_i \end{cases}$$

Therefore, we have

$$\begin{cases} c_i = 1 - \alpha_i - \beta_i \\ d_i = \alpha_i \beta_i \end{cases}, \qquad i = 1, 2, 3, \ldots, n \tag{3.3}$$

Continue with the behavior at infinity and use $z = 1/\zeta$, see Section 2.5. The differential equation in the variable ζ is, see (2.24),

$$\zeta^2 \frac{d^2 u}{d\zeta^2} + \zeta \left(2 - \frac{1}{\zeta} p(1/\zeta) \right) \frac{du}{d\zeta} + \frac{1}{\zeta^2} q(1/\zeta) u = 0$$

and, see Definition 2.5,

$$2 - \frac{1}{\zeta} p(1/\zeta) \text{ and } \frac{1}{\zeta^2} q(1/\zeta), \qquad \text{analytic at } \zeta = 0$$

Proceed as above in the case of a singularity at a finite location, and we get, see (2.14) and (3.1),

$$\alpha + \beta = 1 - \lim_{\zeta \to 0} \left(2 - \frac{1}{\zeta} p(1/\zeta) \right) = -1 + \lim_{\zeta \to 0} \sum_{r=1}^{n} \frac{c_r}{1 - a_r \zeta} = -1 + \sum_{r=1}^{n} c_r$$

With the result in (3.3), we obtain

$$\alpha + \beta + \sum_{r=1}^{n} (\alpha_r + \beta_r) = n - 1$$

Similarly, we get for the product of the two roots, $\alpha\beta$, that the coefficients in the polynomial $R(z) = \sum_{i=0}^{n-2} r_i z^i$ have to satisfy a condition. We assume $n \geq 2$, and we get

$$\alpha\beta = \lim_{\zeta \to 0} \frac{1}{\zeta^2} q(1/\zeta) = \lim_{\zeta \to 0} \left(\sum_{r=1}^{n} \frac{d_r}{(1 - a_r \zeta)^2} + \frac{\zeta^{n-2}}{\zeta^n \psi(1/\zeta)} R(1/\zeta) \right)$$

$$= \sum_{r=1}^{n} d_r + \lim_{\zeta \to 0} \zeta^{n-2} R(1/\zeta) = \sum_{r=1}^{n} d_r + r_{n-2}$$

Finally, a condition on the coefficients f_r is derived.

$$\alpha\beta = \lim_{\zeta \to 0} \frac{1}{\zeta^2} q(1/\zeta) = \lim_{\zeta \to 0} \left(\sum_{r=1}^{n} \frac{d_r}{(1 - a_r \zeta)^2} + \frac{1}{\zeta} \sum_{r=1}^{n} \frac{f_r}{1 - a_r \zeta} \right)$$

which implies

$$\alpha\beta = \sum_{r=1}^{n} d_r + \lim_{\zeta \to 0} \frac{1}{\zeta} \sum_{r=1}^{n} f_r \left(1 + a_r \zeta + O(\zeta^2)\right)$$

For this relation to be consistent, we obtain

$$\alpha\beta = \sum_{r=1}^{n} d_r + \sum_{r=1}^{n} f_r a_r \quad \text{and} \quad \sum_{r=1}^{n} f_r = 0$$

These results are summarized in the following theorem:

Theorem 3.1. *Assume the differential equation of Fuchsian type*

$$\frac{d^2 u(z)}{dz^2} + p(z)\frac{du(z)}{dz} + q(z)u(z) = 0$$

has n distinct, regular singular points in the finite complex plane, located at $z = a_r$, $r = 1, 2, 3, \ldots, n$, and one at $z = \infty$. If the roots of the indicial equation at the regular singular points in the finite plane and at infinity are denoted α_r, β_r, $r = 1, 2, 3, \ldots, n$, and α, β, respectively, the functions $p(z)$ and $q(z)$ of the differential equation have the form

$$\begin{cases} p(z) = \sum_{r=1}^{n} \dfrac{1 - \alpha_r - \beta_r}{z - a_r} \\[2mm] q(z) = \sum_{r=1}^{n} \dfrac{\alpha_r \beta_r}{(z - a_r)^2} + \sum_{r=1}^{n} \dfrac{f_r}{z - a_r} = \sum_{r=1}^{n} \dfrac{\alpha_r \beta_r}{(z - a_r)^2} + \dfrac{1}{\psi(z)} \sum_{i=0}^{n-2} r_i z^i \end{cases}$$

where $\psi(z) = \prod_{r=1}^{n}(z - a_r)$. Moreover, the following consistency relations have to be satisfied

$$\alpha + \beta + \sum_{r=1}^{n} (\alpha_r + \beta_r) = n - 1 \tag{3.4}$$

and

$$\begin{cases} \sum_{r=1}^{n} f_r = 0 \\[2mm] \alpha\beta = \sum_{r=1}^{n} \alpha_r \beta_r + \sum_{r=1}^{n} f_r a_r \end{cases}$$

or

$$\alpha\beta = \sum_{r=1}^{n} \alpha_r \beta_r + r_{n-2} \quad (n \geq 2)$$

Specifically, when $n = 2$, f_1 and $f_2 = -f_1$ are determined in terms of α, β, and α_i, β_i.

3.2 Regular point at infinity

Again, assume there are n distinct, regular singular points in the finite complex plane, located at $z = a_r$, $r = 1, 2, 3, \ldots, n$, but now with a regular point at $z = \infty$. The corresponding roots of the indicial equation are denoted α_r, β_r, $r = 1, 2, 3, \ldots, n$, respectively, and defined as in Section 3.1,

$$p(z) = \frac{G(z)}{\psi(z)}, \qquad q(z) = \frac{H(z)}{(\psi(z))^2}$$

where

$$\psi(z) = \prod_{r=1}^{n} (z - a_r)$$

and where $G(z)$ and $H(z)$ are entire functions such that, see Definition 2.5 on page 24,

$$2z - z^2 p(z) \text{ and } z^4 q(z) \text{ are analytic at } z = \infty$$

With similar arguments as in Section 3.1, these conditions imply that

$$\begin{cases} G(z) \text{ is a polynomial of degree } \leq n - 1 \text{ with coefficient 2 in front of } z^{n-1} \\ H(z) \text{ is a polynomial of degree } \leq 2n - 4 \end{cases}$$

We adopt a similar partial fraction decomposition of the functions $p(z)$ and $q(z)$ as in Section 3.1, viz.,

$$\begin{cases} p(z) = \sum_{r=1}^{n} \frac{c_r}{z - a_r} \\ q(z) = \sum_{r=1}^{n} \frac{d_r}{(z - a_r)^2} + \frac{S(z)}{\psi(z)} = \sum_{r=1}^{n} \frac{d_r}{(z - a_r)^2} + \sum_{r=1}^{n} \frac{f_r}{z - a_r} \end{cases}$$

where $S(z)$ is a polynomial of degree $n - 2$, satisfying two side conditions, see below. The leading coefficients in these representations are determined by

$$\begin{cases} c_i = \lim_{z \to a_i} (z - a_i) p(z) \\ d_i = \lim_{z \to a_i} (z - a_i)^2 q(z) \end{cases} \qquad i = 1, 2, 3, \ldots, n$$

There are conditions on both the $p(z)$ and the $q(z)$ representations. We start with $p(z)$ and investigate the condition imposed by the regularity at infinity.

$$\frac{2}{\zeta} - \frac{1}{\zeta^2} p(1/\zeta) \text{ is analytic at } \zeta = 0$$

or

$$\lim_{\zeta \to 0} \frac{1}{\zeta} \left(2 - \sum_{r=1}^{n} \frac{c_r}{1 - a_r \zeta} \right) = \lim_{\zeta \to 0} \frac{1}{\zeta} \left(2 - \sum_{r=1}^{n} c_r \left(1 + a_r \zeta + O(\zeta^2) \right) \right) \quad \text{finite at } \zeta = 0$$

which implies

$$\sum_{r=1}^{n} c_r = 2 \tag{3.5}$$

We proceed by investigating $q(z)$ and the conditions imposed by the regularity at infinity. There are two conditions on the polynomial $S(z)$.

$$H(z) = q(z) (\psi(z))^2 = (\psi(z))^2 \sum_{r=1}^{n} \frac{d_r}{(z - a_r)^2} + \psi(z) S(z)$$

The degree of the polynomial on the left-hand side is required to be at most $2n - 4$. If we denote the polynomial $S(z) = \sum_{r=0}^{n-2} s_r z^r$, we get, assuming $n \geq 2$, see Problem 3.1,

$$s_{n-2} = - \sum_{r=1}^{n} d_r, \quad s_{n-3} = \left(\sum_{r=1}^{n} d_r \right) \left(\sum_{i=1}^{n} a_i \right) - 2 \sum_{r=1}^{n} d_r a_r \tag{3.6}$$

An alternative representation of the quotient is

$$\frac{S(z)}{\psi(z)} = \sum_{r=1}^{n} \frac{f_r}{z - a_r} \tag{3.7}$$

with **three** conditions on f_r, see Problem 3.2,

$$\sum_{r=1}^{n} f_r = 0, \quad \sum_{r=1}^{n} d_r + \sum_{r=1}^{n} f_r a_r = 0, \quad 2 \sum_{r=1}^{n} d_r a_r + \sum_{r=1}^{n} f_r a_r^2 = 0$$

Analogous to the case with a regular singular point at infinity in Section 3.1, we also have for the roots of the indicial equation, see (3.3),

$$\begin{cases} c_i = 1 - \alpha_i - \beta_i \\ d_i = \alpha_i \beta_i \end{cases}, \quad i = 1, 2, 3, \ldots, n$$

The condition in (3.5) then reads

$$\sum_{r=1}^{n} \alpha_r + \beta_r = \sum_{r=1}^{n} (1 - c_r) = n - 2$$

We summarize these results in the following theorem:

Theorem 3.2. *Assume the differential equation of Fuchsian type*

$$\frac{d^2 u(z)}{dz^2} + p(z) \frac{du(z)}{dz} + q(z) u(z) = 0$$

has n distinct, regular singular points in the finite complex plane, located at $z = a_r$, $r = 1, 2, 3, \ldots, n$, and a regular point at $z = \infty$. If the roots of the indicial equation at the regular singular points are denoted α_r, β_r, $r = 1, 2, 3, \ldots, n$, the functions $p(z)$ and $q(z)$ of the differential equation have the form

$$
\begin{cases}
p(z) = \displaystyle\sum_{r=1}^{n} \frac{1 - \alpha_r - \beta_r}{z - a_r} \\[2ex]
q(z) = \displaystyle\sum_{r=1}^{n} \frac{\alpha_r \beta_r}{(z - a_r)^2} + \sum_{r=1}^{n} \frac{f_r}{z - a_r} = \sum_{r=1}^{n} \frac{\alpha_r \beta_r}{(z - a_r)^2} + \frac{1}{\psi(z)} \sum_{r=0}^{n-2} s_r z^r
\end{cases}
$$

where $\psi(z) = \prod_{r=1}^{n}(z - a_r)$. Moreover, the following consistency relations have to be satisfied

$$
\sum_{r=1}^{n} \alpha_r + \beta_r = n - 2
$$

and

$$
\begin{cases}
\displaystyle\sum_{r=1}^{n} f_r = 0 \\[2ex]
\displaystyle\sum_{r=1}^{n} \alpha_r \beta_r + \sum_{r=1}^{n} f_r a_r = 0 \\[2ex]
2 \displaystyle\sum_{r=1}^{n} \alpha_r \beta_r a_r + \sum_{r=1}^{n} f_r a_r^2 = 0
\end{cases}
$$

or

$$
\begin{cases}
s_{n-2} = -\displaystyle\sum_{r=1}^{n} \alpha_r \beta_r \\[2ex]
s_{n-3} = \left(\displaystyle\sum_{r=1}^{n} \alpha_r \beta_r \right) \left(\sum_{i=1}^{n} a_i \right) - 2 \sum_{r=1}^{n} \alpha_r \beta_r a_r
\end{cases} \qquad (n \geq 2)
$$

Specifically, when $n = 3$, f_1, f_2, and f_3 are determined in terms of α_i and β_i, since[4]

$$
\begin{vmatrix}
1 & 1 & 1 \\
a_1 & a_2 & a_3 \\
a_1^2 & a_2^2 & a_3^2
\end{vmatrix} \neq 0
$$

Comment 3.1. Notice that both the condition above

---

[4] This is a Vandermonde determinant

$$
\begin{vmatrix}
1 & 1 & \ldots & 1 \\
a_1 & a_2 & \ldots & a_n \\
\vdots & \vdots & & \vdots \\
a_1^n & a_2^n & \ldots & a_n^n
\end{vmatrix} = \prod_{i>j}(a_i - a_j)
$$

named after the French musician and chemist Alexandre-Théophile Vandermonde (1735–1796).

$$\sum_{r=1}^{n} \alpha_r + \beta_r = n - 2$$

and the corresponding result of Theorem 3.1, see (3.4),

$$\dot{\alpha} + \beta + \sum_{r=1}^{n} (\alpha_r + \beta_r) = n - 1$$

can be summarized as "the sum of the roots of the indicial equation (including the root of the point at infinity) is equal to the number of regular singular points minus 2." ■

3.3 The displacement theorem

We adopt the same assumptions as in Section 3.1, viz., there are n distinct, regular singular points in the finite complex plane, located at $z = a_r$, $r = 1, 2, 3, \ldots, n$, and one regular singular point at $z = \infty$. The roots of the indicial equation are denoted by α_r, β_r, $r = 1, 2, 3, \ldots, n$, and at infinity they are α, β.

Let $u(z)$ be a solution to (2.1), and introduce a new function $v(z)$ by a multiplication of powers of the following form:

$$u(z) = v(z) \prod_{r=1}^{n} (z - a_r)^{\rho_r} \tag{3.8}$$

where ρ_r, $r = 1, 2, 3, \ldots, n$ are arbitrary complex numbers. This multiplication leads to a change of differential equation, but the location of the singular points and their singular classification remain. The roots of the indicial equation, however, shift or are displaced. The following theorem makes these statements more precise:

Theorem 3.3. *Let $u(z)$ satisfy the differential equation of Fuchsian type*

$$\frac{d^2 u(z)}{dz^2} + p(z) \frac{du(z)}{dz} + q(z) u(z) = 0$$

and assume the equation has n distinct, regular singular points in the finite complex plane, located at $z = a_r$, $r = 1, 2, 3, \ldots, n$, and one regular singular point at $z = \infty$. The roots of the indicial equation at the regular singular points in the finite plane and at infinity are denoted α_r, β_r, $r = 1, 2, 3, \ldots, n$, and α, β, respectively. Moreover, let

$$u(z) = v(z) \prod_{r=1}^{n} (z - a_r)^{\rho_r}$$

where ρ_r, $r = 1, 2, 3, \ldots, n$ are arbitrary complex numbers.
    Then the function $v(z)$ satisfies

$$\frac{d^2v(z)}{dz^2} + \overline{p}(z)\frac{dv(z)}{dz} + \overline{q}(z)v(z) = 0 \tag{3.9}$$

where

$$\begin{cases} \overline{p}(z) = p(z) + \sum_{r=1}^{n}\frac{2\rho_r}{z-a_r} = \sum_{r=1}^{n}\frac{1-\alpha_r-\beta_r+2\rho_r}{z-a_r} = \sum_{r=1}^{n}\frac{1-\overline{\alpha}_r-\overline{\beta}_r}{z-a_r} \\ \overline{q}(z) = q(z) + \frac{d}{dz}\left(\sum_{r=1}^{n}\frac{\rho_r}{z-a_r}\right) + \left(\sum_{r=1}^{n}\frac{\rho_r}{z-a_r}\right)^2 + \left(\sum_{r=1}^{n}\frac{c_r}{z-a_r}\right)\left(\sum_{r=1}^{n}\frac{\rho_r}{z-a_r}\right) \\ \quad\ \ = \sum_{r=1}^{n}\frac{\overline{\alpha}_r\overline{\beta}_r}{(z-a_r)^2} + \frac{\overline{R}(z)}{\psi(z)} \end{cases}$$

which has the same regular singular points, and the indicial equation for $v(z)$ has the roots

$$\begin{cases} \overline{\alpha}_r = \alpha_r - \rho_r \\ \overline{\beta}_r = \beta_r - \rho_r \end{cases} \quad r = 1, 2, 3, \ldots, n \qquad \begin{cases} \overline{\alpha} = \alpha + \sum_{r=1}^{n}\rho_r \\ \overline{\beta} = \beta + \sum_{r=1}^{n}\rho_r \end{cases}$$

corresponding to the finite regular singular points and the point at infinity, respectively.

Proof. Let $u(z) = h(z)v(z)$ and apply Theorem 2.1 on page 4. We have

$$v'' + \left(p+2\frac{h'}{h}\right)v' + \left\{\left(\frac{h'}{h}\right)^2 + \frac{d}{dz}\frac{h'}{h} + p\frac{h'}{h} + q\right\}v = 0$$

From this we conclude that

$$\begin{cases} \overline{p}(z) = p(z) + 2\frac{h'(z)}{h(z)} \\ \overline{q}(z) = \left(\frac{h'(z)}{h(z)}\right)^2 + \frac{d}{dz}\frac{h'(z)}{h(z)} + p(z)\frac{h'(z)}{h(z)} + q(z) \end{cases}$$

We use

$$h(z) = \prod_{r=1}^{n}(z-a_r)^{\rho_r}$$

and get

$$\overline{p}(z) = 2\frac{d}{dz}\ln h(z) + p(z) = 2\frac{d}{dz}\ln\prod_{r=1}^{n}(z-a_r)^{\rho_r} + p(z)$$

which we simplify to

$$\overline{p}(z) = 2\frac{d}{dz}\sum_{r=1}^{n}\ln(z-a_r)^{\rho_r} + p(z) = \sum_{r=1}^{n}\frac{2\rho_r}{z-a_r} + p(z)$$

$$= \sum_{r=1}^{n}\frac{2\rho_r}{z-a_r} + \sum_{r=1}^{n}\frac{c_r}{z-a_r} = \sum_{r=1}^{n}\frac{1-\alpha_r-\beta_r+2\rho_r}{z-a_r}$$

by (3.1) and (3.3). Similarly, we get

$$\overline{q}(z) = \left(\frac{h'(z)}{h(z)}\right)^2 + \frac{d}{dz}\left(\frac{h'(z)}{h(z)}\right) + p(z)\frac{h'(z)}{h(z)} + q(z)$$

$$= \left(\sum_{r=1}^{n}\frac{\rho_r}{z-a_r}\right)^2 + \frac{d}{dz}\left(\sum_{r=1}^{n}\frac{\rho_r}{z-a_r}\right) + \left(\sum_{r=1}^{n}\frac{c_r}{z-a_r}\right)\left(\sum_{r=1}^{n}\frac{\rho_r}{z-a_r}\right) + q(z)$$

and we see that the position and the classification of the singular points a_r, $r = 1, 2, 3, \ldots, n$ remain.

To proceed, calculate the roots of the indicial equations at the singular points. We make the following definitions:

$$\begin{cases} \overline{c}_r = \lim_{z\to a_r}(z-a_r)\overline{p}(z) = 1-\alpha_r-\beta_r+2\rho_r = 1-(\alpha_r-\rho_r)-(\beta_r-\rho_r) \\[2mm] \overline{d}_r = \lim_{z\to a_r}(z-a_r)^2\overline{q}(z) = \rho_r^2-\rho_r+c_r\rho_r+\alpha_r\beta_r \\[2mm] \qquad = \rho_r^2-\rho_r+(1-\alpha_r-\beta_r)\rho_r+\alpha_r\beta_r = \rho_r^2-(\alpha_r+\beta_r)\rho_r+\alpha_r\beta_r \\[2mm] \qquad = (\alpha_r-\rho_r)(\beta_r-\rho_r) \end{cases}$$

since $\lim_{z\to a_r}(z-a_r)^2 q(z) = \alpha_r\beta_r$, and where we also have used (3.3). The indicial equation of the differential equation of $v(z)$ at the regular singular point $z = a_r$, see (2.12), therefore becomes[5]

$$\xi^2 + (\overline{c}_r - 1)\xi + \overline{d}_r = 0$$

or

$$\xi^2 - \xi\left[(\alpha_r-\rho_r)+(\beta_r-\rho_r)\right]+(\alpha_r-\rho_r)(\beta_r-\rho_r) = 0$$

with solutions

$$\begin{cases} \overline{\alpha}_r = \alpha_r - \rho_r \\ \overline{\beta}_r = \beta_r - \rho_r \end{cases}$$

Notice that

$$\overline{p}(z) = \sum_{r=1}^{n}\frac{1-\overline{\alpha}_r-\overline{\beta}_r}{z-a_r}$$

The regular singular point at infinity is analyzed in the same way. The equation transforms with $z = 1/\zeta$ into, see Section 2.5,

---

[5] We denote the variable in the indicial equation by ξ to avoid confusion with the root α to the indicial equation of the regular singular point at infinity.

$$\frac{d^2u}{d\zeta^2} + \frac{1}{\zeta}\left(2 - \frac{1}{\zeta}\overline{p}(1/\zeta)\right)\frac{du}{d\zeta} + \frac{1}{\zeta^4}\overline{q}(1/\zeta)u = 0$$

where

$$\begin{cases} \overline{p}(z) = \sum_{r=1}^{n} \frac{1 - \alpha_r - \beta_r + 2\rho_r}{z - a_r} \\ \overline{q}(z) = q(z) - \sum_{r=1}^{n} \frac{\rho_r}{(z - a_r)^2} + \left(\sum_{r=1}^{n} \frac{\rho_r}{z - a_r}\right)^2 + \left(\sum_{r=1}^{n} \frac{c_r}{z - a_r}\right)\left(\sum_{r=1}^{n} \frac{\rho_r}{z - a_r}\right) \end{cases}$$

To find the roots of the indicial equation of the singularity at infinity, we need to calculate the following limit, cf. Definition 2.5 on page 24:

$$\overline{p}_\infty = \lim_{\zeta \to 0}\left(2 - \frac{1}{\zeta}\overline{p}(1/\zeta)\right) = 2 - \lim_{\zeta \to 0}\sum_{r=1}^{n} \frac{1 - \alpha_r - \beta_r + 2\rho_r}{(1 - a_r\zeta)}$$

$$= 2 - \sum_{r=1}^{n}(1 - \alpha_r - \beta_r + 2\rho_r) = 1 - 2\sum_{r=1}^{n}\rho_r - \alpha - \beta$$

where we in the last equality have used (3.4).

Similarly, we get

$$\overline{q}_\infty = \lim_{\zeta \to 0}\frac{\overline{q}(1/\zeta)}{\zeta^2} = \lim_{\zeta \to 0}\frac{q(1/\zeta)}{\zeta^2} - \lim_{\zeta \to 0}\sum_{r=1}^{n}\frac{\rho_r}{(1 - a_r\zeta)^2}$$

$$+ \lim_{\zeta \to 0}\left(\sum_{r=1}^{n}\frac{\rho_r}{1 - a_r\zeta}\right)^2 + \lim_{\zeta \to 0}\left(\sum_{r=1}^{n}\frac{c_r}{1 - a_r\zeta}\right)\left(\sum_{r=1}^{n}\frac{\rho_r}{1 - a_r\zeta}\right)$$

$$= \alpha\beta - \sum_{r=1}^{n}\rho_r + \left(\sum_{r=1}^{n}\rho_r\right)^2 + \sum_{r=1}^{n}c_r\sum_{r=1}^{n}\rho_r$$

Since $c_r = 1 - \alpha_r - \beta_r$, we get

$$\overline{q}_\infty = \alpha\beta + \left(\sum_{r=1}^{n}\rho_r\right)^2 + \left((n - 1) - \sum_{r=1}^{n}(\alpha_r + \beta_r)\right)\sum_{r=1}^{n}\rho_r$$

and due to (3.4), we get

$$\overline{q}_\infty = \alpha\beta + \left(\sum_{r=1}^{n}\rho_r\right)^2 + (\alpha + \beta)\sum_{r=1}^{n}\rho_r = \left(\alpha + \sum_{r=1}^{n}\rho_r\right)\left(\beta + \sum_{r=1}^{n}\rho_r\right)$$

Since $\overline{p}_\infty$ and $\overline{q}_\infty$ are finite, the regular singular point at infinity remain, see Definition 2.5 on page 24.

The roots of the indicial equation, $\overline{\alpha}, \overline{\beta}$, for this equation at the regular singular point at infinity satisfy, cf. (2.14),

$$\begin{cases} \overline{\alpha} + \overline{\beta} = 1 - \overline{p}_\infty = 2\sum_{r=1}^{n} \rho_r + \alpha + \beta \\[2mm] \overline{\alpha}\overline{\beta} = \overline{q}_\infty = \left(\alpha + \sum_{r=1}^{n} \rho_r \right) \left(\beta + \sum_{r=1}^{n} \rho_r \right) \end{cases}$$

which gives

$$\overline{\alpha} = \alpha + \sum_{r=1}^{n} \rho_r, \qquad \overline{\beta} = \beta + \sum_{r=1}^{n} \rho_r$$

and the theorem is proved. □

As a consequence of this theorem, we see that if we can find solutions to (3.9), then we have a solution to our original equation, (2.1), by (3.8). Notice also that

$$\overline{d}_r = \overline{\alpha}_r \overline{\beta}_r, \qquad r = 1,2,3,\dots,n$$

We conclude this chapter with an example and a comment that points out the alterations in Theorem 3.3 when the point at infinity is a regular point.

Example 3.1. One special case is of interest, viz., $\rho_r = \beta_r$, $r = 1,2,3,\dots,n$, which implies

$$\begin{cases} \overline{\alpha}_r = \alpha_r - \beta_r \\[2mm] \overline{\beta}_r = 0 \end{cases}$$

and the differential equation has the form

$$\frac{d^2 v(z)}{dz^2} + \overline{p}(z)\frac{dv(z)}{dz} + \overline{q}(z)v(z) = 0$$

whereby Theorems 3.1 and 3.2

$$\begin{cases} \overline{p}(z) = \sum_{r=1}^{n} \frac{1 - \overline{\alpha}_r - \overline{\beta}_r}{z - a_r} = \sum_{r=1}^{n} \frac{1 - \overline{\alpha}_r}{z - a_r} \\[3mm] \overline{q}(z) = \sum_{r=1}^{n} \frac{\overline{\alpha}_r \overline{\beta}_r}{(z - a_r)^2} + \frac{\overline{R}(z)}{\psi(z)} = \frac{\overline{R}(z)}{\psi(z)} \end{cases}$$

Notice that the singular behavior of the last term in the differential equation now is altered — it is less singular — but the point is still a regular singular point due to the singularity in $\overline{p}(z)$. ∎

Comment 3.2. Finally, we conclude that if the point, $z = \infty$, is a regular point to the equations, the displacement theorem is unaltered except that we now do not have any displacement of the roots of the indicial equation at infinity. Moreover, if we require the regularity of the point at infinity is preserved, the sum of the shifts must add up to zero, i.e., $\sum_{r=1}^{n} \rho_r = 0$. ∎

Problems

3.1. Show the statement made in (3.6).

3.2. Verify the three conditions on f_r in the representation

$$\frac{S(z)}{\psi(z)} = \sum_{r=1}^{n} \frac{f_r}{z - a_r}$$

in Equation (3.7).

3.3. Show that

$$\bar{\alpha} + \bar{\beta} + \sum_{r=1}^{n} \left(\bar{\alpha}_r + \bar{\beta}_r \right) = n - 1$$

with the notation used in Theorem 3.3.

3.4. Let $u(z)$ satisfy the differential equation

$$\frac{d^2 u(z)}{dz^2} + p(z) \frac{du(z)}{dz} + q(z) u(z) = 0$$

where

$$\begin{cases} p(z) = \dfrac{1 - \alpha_1 - \beta_1}{z} + \dfrac{1 - \alpha_2 - \beta_2}{z - 1} \\ q(z) = -\dfrac{1}{z(z-1)} \left(\dfrac{\alpha_1 \beta_1}{z} - \dfrac{\alpha_2 \beta_2}{z - 1} - a \right) \end{cases}$$

Under what conditions on the constants α_r, β_r, $r = 1, 2$, and a is this a differential equation of Fuchsian type with regular singular points at $z = 0, 1, \infty$? Use the displacement theorem, Theorem 3.3, to transform the differential equation so that one of the roots of the indicial equation at $z = 0, 1$ is zero.

3.5. [†]Let $u(z)$ satisfy

$$\frac{d^2 u(z)}{dz^2} + p(z) \frac{du(z)}{dz} + q(z) u(z) = 0$$

which is assumed to have n distinct, regular singular points in the finite complex plane, located at $z = a_r$, $r = 1, 2, 3, \ldots, n$, and a regular point at $z = \infty$. The roots of the indicial equation are denoted α_r, β_r, $r = 1, 2, 3, \ldots, n$. Prove that a change of the independent variable

$$t(z) = \frac{az + b}{cz + d}, \qquad ad - bc \neq 0$$

transforms the differential equation to a new differential equation having the same form, but with regular singular points at

$$t(a_r) = \frac{a a_r + b}{c a_r + d}, \qquad r = 1, 2, 3, \ldots, n$$

Moreover, the roots of the indicial equation at the regular singular points are identical for the two differential equations, and the consistency relations of the two differential equations have the same form. What happens if one of the singular points, say a_n, satisfies $ca_n + d = 0$?

Chapter 4
Equations with one to four regular singular points

We are now well prepared to illustrate the theory presented in Chapter 3 by some examples. The most simple type of ordinary differential equation contains one regular singular point, either at a finite location in the complex plane or located at infinity. We also present the cases of two, three, and four regular singular points.

Before we determine the form of the ordinary differential equation with one, two, three, and four regular singular points, we might, for completeness, conclude that there does not exist any ordinary differential equation with only regular points in the extended complex plane. This is seen by the conflicting conditions that $p(z)$ and $2z - z^2 p(z)$, and $q(z)$ and $z^4 q(z)$, are all analytic in the extended complex plane, see Definition 2.5 at page 24.

4.1 ODE with one regular singular point

If there is only one regular singular point present, the singular point can be located in the finite complex plane or at infinity. The form of the solutions is similar in the two cases, but we prefer to treat the two cases separately.

4.1.1 Regular singular point at infinity

If the regular singular point is at infinity, we have from Theorem 3.1 on page 32 ($n = 0$) that $p(z) = q(z) = 0$, and

$$u''(z) = 0$$

since at all points in the finite complex plane $p(z)$ and $q(z)$ are analytic, and at the regular singularity at infinity we have

G. Kristensson, *Second Order Differential Equations: Special Functions and Their Classification*, DOI 10.1007/978-1-4419-7020-6_4, © Springer Science+Business Media, LLC 2010

$$zp(z) \text{ and } z^2 q(z) \text{ analytic at } z = \infty$$

We see that there are no free parameters in the differential equation in this case. The solutions are[1]

$$u(z) = Az + B$$

4.1.2 Regular point at infinity

Assume we have only one regular singular point at $z = a_1$ to the ODE ($n = 1$), which we, without loss of generality, can locate at the origin,[2] i.e., $a_1 = 0$. The differential equation is

$$u''(z) + p(z)u'(z) + q(z)u(z) = 0$$

and, in the language of Theorem 3.2 on page 34, the functions must satisfy $\alpha_1 + \beta_1 = -1$, $\alpha_1 \beta_1 = 0$, and $f_1 = 0$. We then have

$$\begin{cases} p(z) = \dfrac{1 - \alpha_1 - \beta_1}{z} = \dfrac{2}{z} \\ q(z) = 0 \end{cases}$$

The general ODE with one regular singular point therefore is

$$z^2 u''(z) + 2zu'(z) = 0$$

and again we see that there are no free parameters in the differential equation. The general solution is[3]

$$u(z) = \frac{A}{z} + B$$

Note that the results presented in the two subsections above, with the regular singular point at the origin and at infinity as two separate cases, can easily be transformed to the other case by the change of variable $z \to 1/z$. With this in mind, the two cases represent the same type of differential equation.

---

[1] Notice that the roots of the indicial equation are 1 and 0, which are consistent with the algebraic structure of the general solution.

[2] Change the variable $z \to z - a_1$.

[3] Notice that the roots of the indicial equation are -1 and 0, which are consistent with the algebraic structure of the general solution.

4.2 ODE with two regular singular points

The two regular singular points can both be located at a finite location in the complex plane or one of them can be located at infinity. In the case of two regular singular points in the finite plane, we also have a possibility to study what happens if the two regular singular points are located very close to each other. This analysis of confluence is studied in Section 4.2.3.

4.2.1 One regular singular point at infinity

We assume one of the regular singular points is located at infinity, and the other regular singular point is located at the origin,[4] i.e., $a_1 = 0$.

$$u''(z) + p(z)u'(z) + q(z)u(z) = 0$$

From Theorem 3.1 we conclude ($n = 1$)

$$\begin{cases} p(z) = \dfrac{1 - \alpha_1 - \beta_1}{z} \\ q(z) = \dfrac{\alpha_1 \beta_1}{z^2} \end{cases}$$

since $f_1 = 0$, and

$$\alpha + \beta + \alpha_1 + \beta_1 = 0$$

where the roots of the indicial equation to the regular singular point at the origin and at infinity are α_1, β_1, and α, β, respectively, and the ordinary differential equation becomes

$$z^2 u''(z) + z(1 - \alpha_1 - \beta_1)u'(z) + \alpha_1 \beta_1 u(z) = 0$$

This equation contains two free parameters, viz., the roots α_1 and β_1 of the indicial equation. The solutions of the differential equation are

$$\begin{cases} u(z) = Az^{\alpha_1} + Bz^{\beta_1}, & \alpha_1 \neq \beta_1 \\ u(z) = z^{\alpha_1}(A + B\ln z), & \alpha_1 = \beta_1 \end{cases}$$

The behavior at infinity is determined by $\zeta = 1/z$. We get

$$\zeta^2 u''(\zeta) + \zeta(1 + \alpha_1 + \beta_1)u'(\zeta) + \alpha_1 \beta_1 u(\zeta) = 0$$

which is a differential equation of the same form as above and with the same roots to the indicial equation, but with opposite sign, i.e., $\alpha = -\alpha_1$ and $\beta = -\beta_1$.

---

[4] Always possible by a change in the variable, see footnote 2.

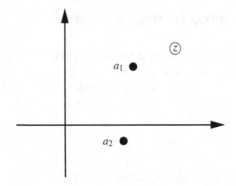

Fig. 4.1 The two regular singular points a_1 and a_2 in the finite complex plane.

4.2.2 Regular point at infinity

If the point at infinity is regular, we use the result of Theorem 3.2 with $n = 2$. The two distinct, regular singular points are $z = a_1$ and $z = a_2$, see Figure 4.1, and the roots of the indicial equation are α_r and β_r, $r = 1, 2$. We have conditions on the roots of the indicial equation

$$
\begin{cases}
p(z) = \displaystyle\sum_{r=1}^{2} \frac{1 - \alpha_r - \beta_r}{z - a_r} \\[2ex]
q(z) = \displaystyle\sum_{r=1}^{2} \frac{\alpha_r \beta_r}{(z - a_r)^2} + \frac{s_0}{(z - a_1)(z - a_2)}
\end{cases}
$$

and constraints

$$
\begin{cases}
s_0 = -\displaystyle\sum_{r=1}^{2} \alpha_r \beta_r \\[2ex]
\displaystyle\sum_{r=1}^{2} (\alpha_r + \beta_r) = 0, \quad \left(\sum_{r=1}^{2} \alpha_r \beta_r \right) \left(\sum_{i=1}^{2} a_i \right) - 2 \sum_{r=1}^{2} \alpha_r \beta_r a_r = 0
\end{cases}
$$

We rewrite the last two conditions as

$$
\begin{cases}
\alpha_2 + \beta_2 = -\alpha_1 - \beta_1 \\
(\alpha_1 \beta_1 - \alpha_2 \beta_2)(a_2 - a_1) = 0 \iff \alpha_1 \beta_1 = \alpha_2 \beta_2
\end{cases}
$$

which implies that $\alpha_1 = -\alpha_2$ and $\beta_1 = -\beta_2$ or $\alpha_1 = -\beta_2$ and $\beta_1 = -\alpha_2$. These conditions lead to

$$
p(z) = \sum_{r=1}^{2} \frac{1 - \alpha_r - \beta_r}{z - a_r} = \frac{1 - \alpha_1 - \beta_1}{z - a_1} + \frac{1 + \alpha_1 + \beta_1}{z - a_2}
$$

and

$$q(z) = \sum_{r=1}^{2} \frac{\alpha_r \beta_r}{(z-a_r)^2} - \sum_{r=1}^{2} \frac{\alpha_r \beta_r}{(z-a_1)(z-a_2)}$$

$$= \frac{\alpha_1 \beta_1}{(z-a_1)^2} + \frac{\alpha_1 \beta_1}{(z-a_2)^2} - \frac{2\alpha_1 \beta_1}{(z-a_1)(z-a_2)}$$

which we simplify to

$$q(z) = \frac{\alpha_1 \beta_1 (a_1 - a_2)^2}{(z-a_1)^2 (z-a_2)^2}$$

and the differential equation becomes

$$u''(z) + \left(\frac{1 - \alpha_1 - \beta_1}{z - a_1} + \frac{1 + \alpha_1 + \beta_1}{z - a_2} \right) u'(z) + \frac{\alpha_1 \beta_1 (a_1 - a_2)^2}{(z-a_1)^2 (z-a_2)^2} u(z) = 0 \quad (4.1)$$

Again, this equation contains two free parameters, viz., the roots α_1 and β_1 of the indicial equation, and the solutions are[5]

$$\begin{cases} u(z) = A \left(\dfrac{z - a_1}{z - a_2} \right)^{\alpha_1} + B \left(\dfrac{z - a_1}{z - a_2} \right)^{\beta_1}, & \alpha_1 \neq \beta_1 \\[3mm] u(z) = \left(\dfrac{z - a_1}{z - a_2} \right)^{\alpha_1} \left(A + B \ln \left(\dfrac{z - a_1}{z - a_2} \right) \right), & \alpha_1 = \beta_1 \end{cases}$$

Note that the results presented in the two subsections above, with the regular singular points at $z = a_1$ and $z = a_2$, and at the origin and at infinity, respectively, as two separate cases, can easily be transformed to the other case by the change of variable $z \rightarrow (z - a_1)/(z - a_2)$. With this in mind, the two cases represent the same type of differential equation.

4.2.3 Confluence

If we have only one regular singular point, we know from Section 4.1 that[6]

$$u''(z) + \frac{2}{z - a} u'(z) = 0 \quad (4.2)$$

with solution

$$u(z) = \frac{A}{z - a} + B$$

On the other hand, if we start with two distinct, regular singular points at $z = a_1$ and $z = a_2$, respectively, and let the two regular singular points coalesce, i.e., $a_1 \rightarrow$

---

[5] Notice that this supports the conclusions $\alpha_1 = -\alpha_2$ and $\beta_1 = -\beta_2$ or $\alpha_1 = -\beta_2$ and $\beta_1 = -\alpha_2$ above.

[6] The regular singular point is shifted from the origin to the point $z = a$.

a_2, we can study what happens with the solution under confluence. To investigate this case, let $a_2 = a_1 + \varepsilon$ in (4.1), and let $\varepsilon \to 0$.

$$u''(z) + \frac{2(z-a_1) - \varepsilon + \varepsilon(\alpha_1 + \beta_1)}{(z-a_1)(z-a_1-\varepsilon)} u'(z) + \frac{\alpha_1 \beta_1 \varepsilon^2}{(z-a_1)^2(z-a_1-\varepsilon)^2} u(z) = 0$$

If we just let $\varepsilon \to 0$ with α_1 and β_1 fixed, we end up with the same equation as above, see (4.2).

To obtain a new equation, we can let the two roots, α_1 and β_1, go to infinity at the same time. Therefore, we impose the constraint that α_1 and β_1 approach infinity according to[7]

$$\begin{cases} \lim_{\varepsilon \to 0} \varepsilon(\alpha_1 + \beta_1) = 0 \\ \lim_{\varepsilon \to 0} \varepsilon^2 \alpha_1 \beta_1 = k^2 \end{cases}$$

where k^2 is a fixed complex constant. Notice that this corresponds to $q_0 \to \infty$, and we expect that classification of the regular singular point at $z = a_1$ changes. We get in the limit $\varepsilon \to 0$

$$u''(z) + \frac{2}{z-a_1} u'(z) + \frac{k^2}{(z-a_1)^4} u(z) = 0$$

Notice that the singular point at $z = a_1$ now is an irregular singular point provided $k^2 \neq 0$, consistent with the fact that $q_0 \to \infty$, and that only one free parameter, k, remains.

To investigate the solution to this equation, let $\zeta = 1/(z-a_1)$, and the equation becomes

$$\frac{d^2 u(\zeta)}{d\zeta^2} + k^2 u(\zeta) = 0$$

with solutions

$$u(z) = A e^{ik/(z-a_1)} + B e^{-ik/(z-a_1)}$$

The fact that the point $z = a_1$ is an irregular singular point also shows up in the functional behavior of the solution at the singular point. We notice that the solution $u(z)$ now has an essential singularity at $z = a_1$, in contrast to the power behavior that always is the singular behavior at a regular singular point.

---

[7] One simple way of accomplishing this is (other possibilities exist)

$$\begin{cases} \alpha_1 = \frac{ik}{\varepsilon} + \frac{f(\varepsilon)}{\varepsilon} \\ \beta_1 = -\frac{ik}{\varepsilon} \end{cases}$$

where $f(\varepsilon)$ is any function satisfying $\lim_{\varepsilon \to 0} f(\varepsilon) = 0$.

4.3 ODE with three regular singular points

We have seen that ordinary differential equations with one or two regular singular points lead to differential equations that have only elementary functions as solutions. Three regular singular points is the simplest type of a differential equation that possesses non-elementary functions as solutions. In this section, we explore this case in some detail. A more extensive treatment is given in Chapter 5.

4.3.1 One regular singular point at infinity

The two regular singular points in the finite plane are, as above, denoted a_1 and a_2, see Figure 4.1. We start by moving these regular singular points to $z = 0, 1$, respectively, and at the same time preserving the regular singular point at infinity.[8] This is accomplished in two steps by changing the variable z. The first change, which moves the point a_1 to the origin, is the change $z \to z - a_1$. The second transformation,[9] $z \to z/a_2$, rotates and stretches the complex plane, which preserves the position of the point at the origin and moves the point a_2 to 1.

Moreover, by the use of the displacement theorem, see Theorem 3.3, we can, without any loss of generality, choose one of the solutions to the indicial equation to zero, say $\beta_1 = \beta_2 = 0$. These transformations show that it suffices to study the case where the regular singular points are located at 0, 1, and ∞, and with the roots of the indicial equation at these points $\{\alpha_1, 0\}$, $\{\alpha_2, 0\}$, and $\{\alpha, \beta\}$, respectively.

To get the form of the ordinary differential equation use Theorem 3.1 ($n = 2$) with $a_1 = 0$ and $a_2 = 1$, and $\beta_1 = \beta_2 = 0$, see also Example 3.1 on page 40. We get

$$
\begin{cases}
p(z) = \dfrac{1-\alpha_1}{z} + \dfrac{1-\alpha_2}{z-1} \\
q(z) = \dfrac{f_1}{z} - \dfrac{f_1}{z-1} = -\dfrac{f_1}{z(z-1)}
\end{cases}
$$

since $f_2 = -f_1$, and subject to the constraints

$$
\begin{cases}
\alpha + \beta + \alpha_1 + \alpha_2 = 1 \\
\alpha\beta = -f_1
\end{cases}
$$

Introduce $\gamma = 1 - \alpha_1$. We then have $\alpha_2 = \gamma - \alpha - \beta$. The general form of the differential equation is therefore

$$
u''(z) + \left[\frac{\gamma}{z} + \frac{1-\gamma+\alpha+\beta}{z-1} \right] u'(z) + \frac{\alpha\beta}{z(z-1)} u(z) = 0
$$

---

[8] The general problem with arbitrary positions of the regular singular points in the finite complex plane is analyzed in Problem 4.2.

[9] The change of variable can, of course, also be made in one step $z \to (z-a_1)/(a_2-a_1)$.

or

$$z(z-1)u''(z) + [(\alpha + \beta + 1)z - \gamma]u'(z) + \alpha\beta u(z) = 0 \qquad (4.3)$$

with regular singular points at $z = 0, 1$, and ∞, and three free parameters α, β, and γ. We notice that the differential equation is uniquely determined by the parameters α, β, and γ, with the convention that the order of α and β is immaterial. This is the **hypergeometric differential equation**, and due to its central role in the solution of many differential equations, it is the subject of a whole chapter, see Chapter 5. We conclude this subsection with a simple lemma that is used below.

Lemma 4.1. *Let $u(z)$ be a solution of the hypergeometric differential equation with parameters α, β, and γ, i.e., (4.3). Then $v(z) = u'(z)$ is the solution to the hypergeometric differential equation with parameters $\alpha + 1$, $\beta + 1$, and $\gamma + 1$ (as usual, the order of the first two parameters is immaterial), i.e.,*

$$z(z-1)v''(z) + [(\alpha + \beta + 3)z - (\gamma + 1)]v'(z) + (\alpha + 1)(\beta + 1)v(z) = 0$$

Proof. Let $u(z)$ be a solution to (4.3), and differentiate the equation. The result is

$$z(z-1)u'''(z) + (2z - 1)u''(z) + [(\alpha + \beta + 1)z - \gamma]u''(z)$$
$$+ (\alpha + \beta + 1)u'(z) + \alpha\beta u'(z) = 0$$

Denote $v(z) = u'(z)$. Then $v(z)$ satisfies

$$z(z-1)v''(z) + [(\alpha + \beta + 3)z - (\gamma + 1)]v'(z) + \underbrace{(\alpha\beta + \alpha + \beta + 1)}_{(\alpha+1)(\beta+1)}v(z) = 0$$

which completes the proof.   □

4.3.2 Regular point at infinity

If the three regular singular points are located in the finite complex plane at a_1, a_2, and a_3,[10] the coefficients of the differential equations are found using the result in Theorem 3.2 with $n = 3$. The result is, see Problem 4.4,

$$\begin{cases} p(z) = \displaystyle\sum_{r=1}^{3} \frac{1 - \alpha_r - \beta_r}{z - a_r} \\[3mm] q(z) = \dfrac{1}{(z - a_1)(z - a_2)(z - a_3)} \displaystyle\sum_{\substack{(ijk) \\ \text{cycl. perm.}}} \dfrac{\alpha_i\beta_i(a_i - a_j)(a_i - a_k)}{z - a_i} \end{cases} \qquad (4.4)$$

where the sum over the cyclic permutations of (ijk) is a sum over the index set $\{ijk\} \in \{(123), (231), (312)\}$. The roots of the indicial equation are subject to the

---

[10] We do not transform these singular points to a set of standard positions as we did in Section 4.3.1.

constraint

$$\sum_{r=1}^{3} \alpha_r + \beta_r = 1$$

The ordinary differential equation with three regular singularities has, because of its widespread use and importance, a special name — the Papperitz[11] equation. Its solutions are denoted

$$u(z) \in P \left\{ \begin{matrix} a_1 & a_2 & a_3 \\ \alpha_1 & \alpha_2 & \alpha_3 & z \\ \beta_1 & \beta_2 & \beta_3 \end{matrix} \right\}$$

This notion is due to Riemann[12] (Riemann's P symbol) and denotes **the set of all solutions** of the differential equation with coefficients (4.4). The first row denotes the three singular points of the differential equation, and the second and third rows contain the roots of the indicial equation, respectively. The three columns can be interchanged arbitrarily. Also, the two roots of the indicial equation in each column can be interchanged, without affecting the solution set.

4.3.2.1 Displacements

The displacement theorem, Theorem 3.3, shows that multiplying the solution with a factor $(z - a_i)^{\rho_i}$ shifts both the roots of the indicial equation by a factor ρ_i, i.e., $\alpha_i \rightarrow \alpha_i - \rho_i$ and $\beta_i \rightarrow \beta_i - \rho_i$. If we want the new solution to be of the same Fuchsian type — in this case with three regular singular points — and preserving the regular point at infinity, the coefficients of the shift must satisfy, see Comment 3.2 on page 40,

$$\sum_{r=1}^{3} \rho_r = 0$$

in order to satisfy the constraint

$$\sum_{r=1}^{3} \alpha_r + \beta_r = 1$$

both for the shifted and the original cases. The result of the displacement theorem then reads

$$P \left\{ \begin{matrix} a_1 & a_2 & a_3 \\ \alpha_1 & \alpha_2 & \alpha_3 & z \\ \beta_1 & \beta_2 & \beta_3 \end{matrix} \right\} = \left(\frac{z - a_1}{z - a_3} \right)^{\rho_1} \left(\frac{z - a_2}{z - a_3} \right)^{\rho_2} P \left\{ \begin{matrix} a_1 & a_2 & a_3 \\ \alpha_1 - \rho_1 & \alpha_2 - \rho_2 & \alpha_3 + \rho_1 + \rho_2 & z \\ \beta_1 - \rho_1 & \beta_2 - \rho_2 & \beta_3 + \rho_1 + \rho_2 \end{matrix} \right\}$$

(4.5)

---

[11] Erwin Johannes Papperitz, German mathematician.
[12] Bernhard Riemann (1826–1866), German mathematician.

4.3.2.2 Substitutions

It makes sense to ask what transformations of the independent variable z preserve the form of the equation — in this case three distinct, regular singular points. Such transformation must be one-to-one mappings of the extended complex plane onto itself. The Möbius[13] transformation $t(z)$ is the only such transformation,[14] and the following theorem is central:

Theorem 4.1 (The substitution theorem). *A change of the independent variable*

$$t(z) = \frac{az+b}{cz+d}, \qquad ad - bc \neq 0$$

preserves the set of solutions, i.e.,

$$P \left\{ \begin{matrix} a_1 & a_2 & a_3 \\ \alpha_1 & \alpha_2 & \alpha_3 & z \\ \beta_1 & \beta_2 & \beta_3 \end{matrix} \right\} = P \left\{ \begin{matrix} t(a_1) & t(a_2) & t(a_3) \\ \alpha_1 & \alpha_2 & \alpha_3 & t(z) \\ \beta_1 & \beta_2 & \beta_3 \end{matrix} \right\}$$

where $\sum_{r=1}^{3} (\alpha_r + \beta_r) = 1$ and the images of the regular singular points are

$$t(a_r) = \frac{aa_r + b}{ca_r + d}, \quad r = 1,2,3$$

If one of the regular singular points, say a_3, satisfies $ca_3 + d = 0$, then the images are two regular singular points, $t(a_1)$ and $t(a_2)$, in the finite complex plane, and a regular singular point at infinity.

Proof. The proof is a direct consequence of the results in Problem 3.5, but we prefer to present an explicit proof using Theorem 2.1 on page 4. To this end, assume $u(z)$ is a solution of

$$\frac{d^2 u(z)}{dz^2} + p(z)\frac{du(z)}{dz} + q(z)u(z) = 0$$

where the coefficients are given by (4.4). Then if

$$t(z) = \frac{az+b}{cz+d}, \text{ with inverse } z(t) = \frac{b - dt}{ct - a}$$

$v(t) = u(z(t))$ satisfies

$$\frac{d^2 v(t)}{dt^2} + \left(p(z(t))z'(t) - \frac{z''(t)}{z'(t)} \right) \frac{dv(t)}{dt} + q(z(t)) \left(z'(t) \right)^2 v(t) = 0$$

In our special case

$$z'(t) = \frac{ad - bc}{(ct - a)^2} \quad \text{and} \quad z''(t) = -2c\frac{ad - bc}{(ct - a)^3}$$

Explicit calculations show that the first coefficient in (4.4) transforms as (we introduce the notion $c_r = 1 - \alpha_r - \beta_r$, $r = 1, 2, 3$)

$$\tilde{p}(t) = p(z(t))z'(t) - \frac{z''(t)}{z'(t)} = \sum_{r=1}^{3} \frac{c_r(ad - bc)}{(b - dt)(ct - a) - a_r(ct - a)^2} + \frac{2c}{ct - a}$$

$$= \frac{1}{(ct - a)} \sum_{r=1}^{3} \left(\frac{c_r(ad - bc)}{(aa_r + b) - t(ca_r + d)} + c_r c \right)$$

$$= \frac{1}{(ct - a)} \sum_{r=1}^{3} c_r \frac{(ad - bc) + c(b + aa_r) - ct(ca_r + d)}{(aa_r + b) - t(ca_r + d)}$$

$$= \frac{1}{(ct - a)} \sum_{r=1}^{3} c_r \frac{(ca_r + d)(ct - a)}{t(ca_r + d) - (aa_r + b)} = \sum_{r=1}^{3} \frac{c_r}{t - t(a_r)}$$

since $\sum_{r=1}^{3} c_r = 2$, and where

$$t(a_r) = \frac{aa_r + b}{ca_r + d}$$

If one of the regular singular points, say a_3, satisfies $ca_3 + d = 0$, then only two terms remain in the sum, i.e.,

$$\tilde{p}(t) = \sum_{r=1}^{2} \frac{c_r}{t - t(a_r)}$$

The second coefficient in (4.4) transforms as

$$\tilde{q}(t) = q(z(t))\left(z'(t)\right)^2$$

$$= \frac{(ad - bc)^2}{[b - dt - a_1(ct - a)][b - dt - a_2(ct - a)][b - dt - a_3(ct - a)]}$$

$$\times \sum_{\substack{(ijk) \\ \text{cycl. perm.}}} \frac{\alpha_i \beta_i (a_i - a_j)(a_i - a_k)}{b - dt - a_i(ct - a)}$$

which we simplify to

$$\tilde{q}(t) = \frac{(ad - bc)^2(ca_1 + d)^{-1}(ca_2 + d)^{-1}(ca_3 + d)^{-1}}{(t - t(a_1))(t - t(a_2))(t - t(a_3))} \sum_{\substack{(ijk) \\ \text{cycl. perm.}}} \frac{\alpha_i \beta_i (a_i - a_j)(a_i - a_k)}{t(ca_i + d) - (aa_i + b)}$$

$$= \frac{(ad - bc)^2}{(t - t(a_1))(t - t(a_2))(t - t(a_3))} \sum_{\substack{(ijk) \\ \text{cycl. perm.}}} \frac{\alpha_i \beta_i (a_i - a_j)(a_i - a_k)}{(t - t(a_i))(ca_i + d)^2(ca_j + d)(ca_k + d)}$$

For each cyclic permutation of i, j, k, the terms in the sum can be rewritten as

$$\frac{(ad-bc)^2(a_i-a_j)(a_i-a_k)}{(ca_i+d)^2(ca_j+d)(ca_k+d)} = \{(aa_i+b)(ca_j+d)-(aa_j+b)(ca_i+d)\}$$

$$\times \frac{\{(aa_i+b)(ca_k+d)-(aa_k+b)(ca_i+d)\}}{(ca_i+d)^2(ca_j+d)(ca_k+d)} = (t(a_i)-t(a_j))(t(a_i)-t(a_k))$$

The entire contribution then is

$$\tilde{q}(t) = \frac{1}{(t-t(a_1))(t-t(a_2))(t-t(a_3))} \sum_{\substack{(ijk)\\ \text{cycl. perm.}}} \frac{\alpha_i\beta_i(t(a_i)-t(a_j))(t(a_i)-t(a_k))}{t-t(a_i)}$$

If one of the regular singular points, say a_3, satisfies $ca_3+d=0$, then we obtain the second coefficient from above by taking the limit $t(a_3) \to \infty$, i.e.,

$$\tilde{q}(t) = \frac{\alpha_1\beta_1(t(a_1)-t(a_2))}{(t-t(a_1))^2(t-t(a_2))} + \frac{\alpha_2\beta_2(t(a_2)-t(a_1))}{(t-t(a_1))(t-t(a_2))^2} + \frac{\alpha_3\beta_3}{(t-t(a_1))(t-t(a_2))}$$

We therefore conclude that, if $u(z)$ is a solution to

$$\frac{d^2u(z)}{dz^2} + p(z)\frac{du(z)}{dz} + q(z)u(z) = 0 \tag{4.6}$$

with coefficients (4.4), then $v(t) = u(z(t))$ satisfies

$$\frac{d^2v(t)}{dt^2} + \tilde{p}(t)\frac{dv(t)}{dt} + \tilde{q}(t)v(t) = 0 \tag{4.7}$$

where

$$\begin{cases} \tilde{p}(t) = \sum_{r=1}^{3} \frac{c_r}{t-t(a_r)} \\[2ex] \tilde{q}(t) = \frac{1}{(t-t(a_1))(t-t(a_2))(t-t(a_3))} \sum_{\substack{(ijk)\\ \text{cycl. perm.}}} \frac{\alpha_i\beta_i(t(a_i)-t(a_j))(t(a_i)-t(a_k))}{t-t(a_i)} \end{cases}$$

The roots of the indicial equation for these two differential equations, (4.6) and (4.7), are the same.

It remains to investigate what happens if one of the regular singular points, say a_3, satisfies $ca_3+d=0$. We then have

$$\begin{cases} \tilde{p}(t) = \sum_{r=1}^{2} \frac{c_r}{t-t(a_r)} \\[2ex] \tilde{q}(t) = \frac{\alpha_1\beta_1(t(a_1)-t(a_2))}{(t-t(a_1))^2(t-t(a_2))} + \frac{\alpha_2\beta_2(t(a_2)-t(a_1))}{(t-t(a_1))(t-t(a_2))^2} + \frac{\alpha_3\beta_3}{(t-t(a_1))(t-t(a_2))} \end{cases}$$

From Definition 2.5 on page 24, we conclude that the point at infinity is a regular singular point, since

$$2 - \lim_{\zeta \to 0} \frac{1}{\zeta} \tilde{p}(1/\zeta) = 2 - \sum_{r=1}^{2} c_r = c_3 = 1 - \alpha_3 - \beta_3$$

and

$$\lim_{\zeta \to 0} \frac{1}{\zeta^2} \tilde{q}(1/\zeta) = \alpha_3 \beta_3$$

are analytic at $\gamma = 0$, and the theorem is proved. □

4.3.2.3 Connection to a regular singular point at infinity

Theorem 4.1 states that the solution set of the two equations, (4.6) and (4.7), are the same. Using this theorem, it is possible to transform one of the singular points to infinity (transforming the infinity point from a regular point to a regular singular point), and, moreover, to scale the other two to the positions $z = 0$ and $z = 1$, respectively. It is convenient to extend the use of the Riemann's P symbol to accommodate also solutions, which have a regular singular point at infinity and two regular singular points at $z = 0$ and $z = 1$. With such an extension, we get

$$P \left\{ \begin{matrix} a_1 & a_2 & a_3 \\ \alpha_1 & \alpha_2 & \alpha & z \\ \beta_1 & \beta_2 & \beta \end{matrix} \right\} = P \left\{ \begin{matrix} 0 & 1 & \infty \\ \alpha_1 & \alpha_2 & \alpha & \frac{z-a_1}{a_2-a_1} \frac{a_2-a_3}{z-a_3} \\ \beta_1 & \beta_2 & \beta \end{matrix} \right\} \tag{4.8}$$

Please bear in mind that the roots of the indicial equation satisfy

$$\sum_{r=1}^{2} \alpha_r + \beta_r + \alpha + \beta = 1$$

which is consistent both with the condition for three regular singular points in the finite complex plane and a regular point at infinity (left-hand side of (4.8)), and with the condition for two regular singular points in the finite complex plane and a regular singular point at infinity (right-hand side (4.8)).

Moreover, with the use of the displacement theorem, Theorem 3.3, and Comment 3.2 on page 40, we can shift one of the roots of the indicial equation at the points at $z = 0$ and $z = 1$ to zero by a shift $\rho_r = \beta_r$. However, to preserve the Fuchsian type — two regular singular points at $z = 0$ and $z = 1$, and one regular singular point at infinity — the root of the indicial equation of the point at infinity has to increase by $\rho_1 + \rho_2 = \beta_1 + \beta_2$. The result of this transformation is

$$P \left\{ \begin{matrix} 0 & 1 & \infty \\ \alpha_1 & \alpha_2 & \alpha & z \\ \beta_1 & \beta_2 & \beta \end{matrix} \right\} = z^{\beta_1} (z-1)^{\beta_2} P \left\{ \begin{matrix} 0 & 1 & \infty \\ 0 & 0 & \alpha + \beta_1 + \beta_2 & z \\ \alpha_1 - \beta_1 & \alpha_2 - \beta_2 & \beta + \beta_1 + \beta_2 \end{matrix} \right\} \tag{4.9}$$

From these last relations, (4.8) and (4.9), we see that the total set of solutions of the Papperitz equation, where all three regular singular points lie in the finite complex plane, can be obtained from the solutions of the differential equation with regular singular points at 0, 1, and ∞, respectively, and where two of the roots of the indicial equation are zero, i.e., the hypergeometric differential equation, (4.3). All solutions of the Papperitz equation are therefore found by solving the hypergeometric equation followed by a proper transformation of the independent variable. This result strongly motivates us to study the solutions of the hypergeometric equation, and this is the subject of Chapter 5.

4.4 ODE with four regular singular points

We proceed with the two different cases of differential equations with four regular singular points. The point at infinity can, as above, be a regular singular point or a regular point. These two cases are developed in the two sections below. A more extensive analysis of the solutions is given in Chapter 8.

4.4.1 One regular singular point at infinity

The three singular points in the finite complex plane are generally located at $z = a_i$, $i = 1, 2, 3$. One of these points can be transformed to the origin, and one transformed (rotated and translated) to $z = 1$, see the analogous transformation and rotation in Section 4.3.1. The remaining singular point is located at $z = a$. The three singular points in the finite complex plane are therefore located at $a_1 = 0$, $a_2 = 1$, and $a_3 = a$ (complex number), see Figure 4.2.

One of the roots of the indicial equation can, for each a_i, by the displacement theorem on page 36, be set to zero, see also Example 3.1 on page 40. The other roots are labeled α_i, $i = 1, 2, 3$, corresponding to the singular points a_i, $i = 1, 2, 3$, respectively. The roots at infinity are denoted α and β.

Following the result of Theorem 3.1 $(n = 3)$ on page 32, we get

$$\begin{cases} p(z) = \dfrac{1 - \alpha_1}{z} + \dfrac{1 - \alpha_2}{z - 1} + \dfrac{1 - \alpha_3}{z - a} \\ q(z) = \dfrac{r_0 + r_1 z}{z(z - 1)(z - a)} \end{cases}$$

subject to the constraints

$$\begin{cases} \alpha + \beta + \alpha_1 + \alpha_2 + \alpha_3 = 2 \\ \alpha\beta = r_1 \end{cases}$$

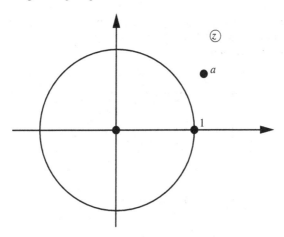

Fig. 4.2 The three regular singular points at $z = 0, 1, a$ in Heun's equation. The fourth singular point is located at infinity.

Introduce

$$\begin{cases} \gamma = 1 - \alpha_1 \\ \delta = 1 - \alpha_2 \\ \varepsilon = 1 - \alpha_3 \end{cases}$$

The general form of the differential equation is therefore

$$u''(z) + \left[\frac{\gamma}{z} + \frac{\delta}{z-1} + \frac{\varepsilon}{z-a} \right] u'(z) + \frac{\alpha\beta(z-h)}{z(z-1)(z-a)} u(z) = 0 \qquad (4.10)$$

where h is a free parameter — an accessory parameter — and the other coefficients are subject to the constraint

$$\alpha + \beta + 1 = \gamma + \delta + \varepsilon$$

This is Heun's equation,[15] and the properties of its solution are analyzed in some detail in Chapter 8.

4.4.2 Regular point at infinity

For completeness, we give the form of the differential equation with four regular singular points in the finite complex plane, and the point at infinity being a regular point, even if we do not explore this equation any further in this textbook.

---

[15] Karl Heun (1859–1929), German mathematician.

Labeling the regular singular points, as usual, by a_i, $i = 1,2,3,4$, and the corresponding roots of the indicial equation by α_i and β_i, $i = 1,2,3,4$, respectively, Theorem 3.2 on page 34 implies

$$
\begin{cases}
p(z) = \displaystyle\sum_{r=1}^{4} \frac{1 - \alpha_r - \beta_r}{z - a_r} \\[3mm]
q(z) = \displaystyle\sum_{r=1}^{4} \frac{\alpha_r \beta_r}{(z - a_r)^2} + \sum_{r=1}^{4} \frac{f_r}{z - a_r}
\end{cases}
$$

subject to

$$
\sum_{r=1}^{4} (\alpha_r + \beta_r) = 2
$$

and

$$
\sum_{r=1}^{4} f_r = 0, \qquad \sum_{r=1}^{4} \alpha_r \beta_r + \sum_{r=1}^{4} f_r a_r = 0, \qquad 2\sum_{r=1}^{4} \alpha_r \beta_r a_r + \sum_{r=1}^{4} f_r a_r^2 = 0
$$

Generalizing the notation by Riemann introduced in Section 4.3.2, we write the set of solutions to the differential equation in this section as

$$
u(z) \in P \begin{Bmatrix} a_1 & a_2 & a_3 & a_4 & \\ \alpha_1 & \alpha_2 & \alpha_3 & \alpha_4 & z \\ \beta_1 & \beta_2 & \beta_3 & \beta_4 & \end{Bmatrix} \tag{4.11}
$$

This notion denotes **the set of all solutions** of the differential equation with coefficients above. The first row denotes the four singular points of the differential equation, and the second and third rows contain the roots of the indicial equation, respectively. The four columns can be interchanged arbitrarily. Also, the two roots of the indicial equation in each column can be interchanged, without affecting the solution set.

Problems

4.1. Prove that the differential equation with two regular singular points, (4.1), transforms by a change of the independent variable

$$
t(z) = \frac{az + b}{cz + d}, \qquad ad - bc \neq 0
$$

to the differential equation

$$
v''(t) + \left(\frac{1 - \alpha_1 - \beta_1}{t - t(a_1)} + \frac{1 + \alpha_1 + \beta_1}{t - t(a_2)} \right) v'(t) + \frac{\alpha_1 \beta_1 (t(a_1) - t(a_2))^2}{(t - t(a_1))^2 (t - t(a_2))^2} v(t) = 0
$$

i.e., the change of variable preserves the set of solutions, cf. Theorem 4.1. What happens if one of the singular points, say a_2, satisfies $ca_2 + d = 0$?

4.2. Find the general form of the differential equation with two regular singular points at $z = a_1$ and $z = a_2$ and one regular singular point at infinity. Check, by specializing the result, with the expression in (4.3).

4.3. Show that the general differential equation with three regular singular points at $z = a_1, a_2$ and ∞ transforms into the equation in Section 4.2.1 as $a_2 \to a_1$.
Hint: Start with the hypergeometric differential equation, see (4.3),

$$z(z-1)u''(z) + [(\alpha + \beta + 1)z - \gamma]u'(z) + \alpha\beta u(z) = 0$$

and make the transformation $z \to (z - a_1)/(a_2 - a_1)$.

4.4. Show the representation of the coefficient functions $p(z)$ and $q(z)$ in the Papperitz equation, see (4.4),

$$\begin{cases} p(z) = \sum_{r=1}^{3} \frac{1 - \alpha_r - \beta_r}{z - a_r} \\ q(z) = \frac{1}{(z - a_1)(z - a_2)(z - a_3)} \sum_{\substack{(ijk) \\ \text{cycl. perm.}}} \frac{\alpha_i \beta_i (a_i - a_j)(a_i - a_k)}{z - a_i} \end{cases}$$

where the sum over the cyclic permutations of (ijk) is a sum over the index set $\{ijk\} \in \{(123), (231), (312)\}$.

4.5. Using the notion in (4.11) and the displacement theorem, Theorem 3.3, prove (also compare (4.5) and (4.9))

$$P\left\{ \begin{matrix} a_1 & a_2 & a_3 & a_4 \\ \alpha_1 & \alpha_2 & \alpha_3 & \alpha_4 & z \\ \beta_1 & \beta_2 & \beta_3 & \beta_4 \end{matrix} \right\} = \left(\frac{z - a_1}{z - a_4} \right)^{\rho_1} \left(\frac{z - a_2}{z - a_4} \right)^{\rho_2} \left(\frac{z - a_3}{z - a_4} \right)^{\rho_3}$$

$$\times P\left\{ \begin{matrix} a_1 & a_2 & a_3 & a_4 \\ \alpha_1 - \rho_1 & \alpha_2 - \rho_2 & \alpha_3 - \rho_3 & \alpha_4 + \rho_1 + \rho_2 + \rho_3 & z \\ \beta_1 - \rho_1 & \beta_2 - \rho_2 & \beta_3 - \rho_3 & \beta_4 + \rho_1 + \rho_2 + \rho_3 \end{matrix} \right\}$$

and

$$P\left\{ \begin{matrix} 0 & 1 & a & \infty \\ \alpha_1 & \alpha_2 & \alpha_3 & \alpha & z \\ \beta_1 & \beta_2 & \beta_3 & \beta \end{matrix} \right\} = z^{\beta_1} (z-1)^{\beta_2} (z-a)^{\beta_3}$$

$$\times P\left\{ \begin{matrix} 0 & 1 & a & \infty \\ 0 & 0 & 0 & \beta + \beta_1 + \beta_2 + \beta_3 & z \\ \alpha_1 - \beta_1 & \alpha_2 - \beta_2 & \alpha_3 - \beta_3 & \alpha + \beta_1 + \beta_2 + \beta_3 \end{matrix} \right\}$$

Chapter 5
The hypergeometric differential equation

Differential equations with three regular singular points (one located at infinity) play an important role in mathematical physics, and in this chapter we investigate this situation in some detail. More details are found in the rich literature on the subject, see, e.g., Refs. 11, 18, 31.

5.1 Basic properties

We have already seen that the hypergeometric differential equation in (4.3)

$$z(z-1)u''(z) + [(\alpha+\beta+1)z - \gamma]u'(z) + \alpha\beta u(z) = 0$$

has three regular singular points at $z = 0, 1, \infty$. We also write the differential equation as

$$u''(z) + \frac{(\alpha+\beta+1)z - \gamma}{z(z-1)}u'(z) + \frac{\alpha\beta}{z(z-1)}u(z) = 0$$

The roots of the indicial equation are given by, see (2.12),

$$I(\lambda) = \lambda^2 + (p_0 - 1)\lambda + q_0 = 0$$

The roots of the indicial equations at the three different regular singular points are summarized in Table 5.1.

Table 5.1 The roots of the indicial equation of the hypergeometric differential equation.

Point	p_0	q_0	Roots λ
$z = 0$	γ	0	$0, 1-\gamma$
$z = 1$	$1+\alpha+\beta-\gamma$	0	$0, \gamma-\alpha-\beta$
$z = \infty$	$1-\alpha-\beta$	$\alpha\beta$	α, β

G. Kristensson, *Second Order Differential Equations: Special Functions and Their Classification*, DOI 10.1007/978-1-4419-7020-6_5,
© Springer Science+Business Media, LLC 2010

We now investigate the solution of the hypergeometric differential equation in the vicinity of the regular singular point $z = 0$, where the roots of the indicial equation are 0 and $1 - \gamma$. Therefore, we expect that one of the solutions is analytic at $z = 0$. In general, from the analysis in Section 2.4, there are two linearly independent solutions $u(z)$ and $v(z)$ of the form

$$u(z) = 1 + \sum_{n=1}^{\infty} a_n z^n, \qquad v(z) = z^{1-\gamma} \left(1 + \sum_{n=1}^{\infty} a_n' z^n \right) \tag{5.1}$$

Note that the solution $v(z)$ is not analytic near the regular singular point $z = 0$, unless $\gamma = 1, 0, -1, -2, \ldots$. However, in this case the two indicial roots differ by an integer, and $\mathrm{Re}(1 - \gamma) \geq 0$, which implies that the results of Section 2.4.3 have to be used. The power series solution $u(z)$ in (5.1) is called the **hypergeometric series**. The radius of convergence of this power series is at most the distance to the closest singular point,[1] i.e., the radius is less than or equal to 1. The properties of this power series solution are presented in Theorem 5.1 below.

The analytic extension of the hypergeometric series in the complex z-plane is denoted the **hypergeometric function**,[2] and we adopt the notation $F(\alpha, \beta; \gamma; z)$. Note that

$$F(\alpha, \beta; \gamma; 0) = 1$$

The hypergeometric function is analytic everywhere in the finite complex plane, excluding the possible singular point at $z = 1$. The singularity can be either a pole or a branch point. If the singularity is a branch point, we introduce a branch cut from $z = 1$ to $z = \infty$ along the real axis, in order to get a one-valued function throughout the cut plane.

A long list of elementary and special functions can be expressed in a direct or an indirect way in the hypergeometric function. A collection of functions that can be expressed in the hypergeometric function is found in Appendix E on page 209.

5.2 Hypergeometric series

The objective of our investigation is now to explicitly determine the coefficients a_n and a_n' in (5.1). This approach resembles strongly the analysis in Section 2.4, i.e., Frobenius method, but due to the special form of the differential equation, we repeat parts of the analysis here.

We begin with the a_n coefficients and insert the power series of $u(z)$ in the differential equation, and identity the coefficient in front of the same power of z, which must vanish in order to have the power series of $u(z)$ satisfying the differential equa-

---

[1] An exception occurs if $\alpha\beta = 0$, then $u(z) = 1$ is a solution, which has infinite convergence radius.

[2] The more complete notion is $_2F_1(\alpha, \beta; \gamma; z)$, see also the generalized hypergeometric series in Equation (7.14) on page 138.

tion.[3] We get

$$n(n-1)a_n - (n+1)na_{n+1} + (\alpha+\beta+1)na_n - \gamma(n+1)a_{n+1} + \alpha\beta a_n = 0, \quad n \in \mathbb{N}$$

which simplifies to

$$(n+1)(n+\gamma)a_{n+1} = (n+\alpha)(n+\beta)a_n, \quad n \in \mathbb{N} \tag{5.2}$$

or

$$\frac{a_{n+1}}{a_n} = \frac{(n+\alpha)(n+\beta)}{(n+1)(n+\gamma)}, \quad n \in \mathbb{N}$$

The values of $\gamma = 0, -1, -2, \ldots$ have to be excluded from the analysis, otherwise the coefficients a_n become undetermined for high n values. By recursion we get ($a_0 = 1$)

$$a_n = \frac{a_n}{a_0} = \prod_{v=0}^{n-1} \frac{a_{v+1}}{a_v} = \prod_{v=0}^{n-1} \frac{(v+\alpha)(v+\beta)}{(v+1)(v+\gamma)}$$

Notice that we can always let $a_0 = 1$, since the hypergeometric differential equation is linear and homogeneous.[4] It is now convenient to introduce the rising factorial, also known as the Appell[5] symbol or Pochhammer[6] symbol, (α, n) defined in Appendix A on page 173 as

$$(\alpha, n) = \prod_{v=0}^{n-1}(v+\alpha) = \alpha(\alpha+1)(\alpha+2)\ldots(\alpha+n-1) = \frac{\Gamma(\alpha+n)}{\Gamma(\alpha)}, \qquad (\alpha, 0) = 1$$

and we write the coefficient a_n as

$$a_n = \frac{(\alpha, n)(\beta, n)}{(\gamma, n)(1, n)} = \frac{(\alpha, n)(\beta, n)}{(\gamma, n)n!}$$

since $n! = (1, n)$. The solution $u(z)$ then becomes

$$u(z) = \sum_{n=0}^{\infty} \frac{(\alpha, n)(\beta, n)}{(\gamma, n)} \frac{z^n}{n!}$$

The sum on the right-hand side defines the hypergeometric series, and the sum is a representation of the hypergeometric function in the domain of the complex plane, where the series converges. The domain of convergence of the hypergeometric series is investigated in the following theorem:

Theorem 5.1. *The hypergeometric series*

---

[3] The coefficients a'_n in (5.1) are determined in Problem 5.1.
[4] When $\gamma = 0$ and $\alpha\beta \neq 0$, the recursion relation for the lowest order term $n = 0$ is $\gamma a_1 = \alpha\beta a_0$, with is in conflict with $a_0 = 1$. In this case, no solution, that is analytic at the origin satisfying $u(0) \neq 0$, exists.
[5] Paul Appell (1855–1930), French mathematician.
[6] Leo August Pochhammer (1841–1920), Prussian mathematician.

$$\sum_{n=0}^{\infty} \frac{(\alpha,n)(\beta,n)}{(\gamma,n)} \frac{z^n}{n!}$$

converges inside the unit circle, $|z| < 1$, and there it represents an analytic function for any choice of parameters, provided γ is not zero or a negative integer.[7] *Outside the unit circle, $|z| > 1$ the series, in general, diverges. On the unit circle, $|z| = 1$, the series is absolutely convergent, provided $\mathrm{Re}(\gamma - \alpha - \beta) > 0$. The series converges, but not absolutely, for $|z| = 1$, $z \neq 1$, as long as $-1 < \mathrm{Re}(\gamma - \alpha - \beta) \leq 0$. If $\mathrm{Re}(\gamma - \alpha - \beta) \leq -1$ the series diverges on $|z| = 1$.*

The hypergeometric series represents the hypergeometric function in the domain of convergence of the series, i.e.,

$$F(\alpha,\beta;\gamma;z) = \sum_{n=0}^{\infty} \frac{(\alpha,n)(\beta,n)}{(\gamma,n)} \frac{z^n}{n!}, \qquad |z| < 1$$

If α or β is a non-positive integer $-n$, the series is a polynomial of degree n.

Proof. The radius of convergence, r, of the power series is determined by the ratio test[8]

$$r^{-1} = \lim_{n \to \infty} \left| \frac{a_{n+1}}{a_n} \right|$$

The analysis preceding this theorem shows that

$$\lim_{n \to \infty} \left| \frac{a_{n+1}}{a_n} \right| = \lim_{n \to \infty} \left| \frac{(n+\alpha)(n+\beta)}{(n+1)(n+\gamma)} \right| = 1$$

and since the radius of convergence, r, of a power series is determined by reciprocal of this limit, we have proved that the power series converges absolutely for $|z| < 1$ and diverges for $|z| > 1$. Convergence inside and divergence outside the unit circle are therefore proved. On the radius of convergence, $r = 1$, the ratio test is inconclusive, and we can use Raabe's test.[9] From above, we get

---

[7] When γ is zero or a negative integer, the γ parameter is increased by an arbitrary integer m if we investigate the mth derivative of $u(z)$ instead, see Lemma 4.1 on page 50 (see also footnote 4). The solution $u(z)$ is then obtained by an m-fold integration.

[8] Sometimes this test is known as d'Alembert's ratio test after the French mathematician Jean le Rond d'Alembert (1717–1783).

[9] Raabe's test, see, e.g., [24] or [13, p. 106], is sometimes useful when the ratio test is inconclusive. The test is named after the Swiss mathematician Joseph Ludwig Raabe (1801–1859). If the terms $w_n \in \mathbb{C}$ of the series

$$S = \sum_{n=0}^{\infty} w_n$$

satisfy

$$\frac{w_{n+1}}{w_n} = 1 + \frac{\alpha}{n} + \frac{\beta_n}{n^2}, \quad |\beta_n| \leq M$$

for large n, then the series S converges absolutely if $\mathrm{Re}\,\alpha < -1$.

$$\frac{a_{n+1}}{a_n} = \frac{(n+\alpha)(n+\beta)}{(n+1)(n+\gamma)} = 1 + \frac{\alpha+\beta-\gamma-1}{n} + O(n^{-2})$$

and by Raabe's test, the power series converges absolutely on the unit circle, $z = 1$, provided $\mathrm{Re}(\gamma - \alpha - \beta) > 0$.

In order to investigate what happens on the unit circle in more detail, let $z = e^{i\theta}$, $\theta \in [0, 2\pi]$. The sum to analyze is

$$\sum_{n=0}^{\infty} a_n e^{in\theta}$$

where

$$a_n = \frac{(\alpha,n)(\beta,n)}{(\gamma,n)n!} = \frac{(\alpha,n)}{n!}\frac{(\beta,n)}{n!}\frac{n!}{(\gamma,n)} = \frac{\Gamma(\gamma)}{\Gamma(\alpha)\Gamma(\beta)} n^{\alpha+\beta-\gamma-1}(1+O(1/n))$$

$$= \frac{\Gamma(\gamma)}{\Gamma(\alpha)\Gamma(\beta)}\left(n^{-(\delta+1)} + O(n^{-(\mathrm{Re}\,\delta+2)})\right)$$

where we used Lemma A.5 on page 174, and where $\delta = \gamma - \alpha - \beta$. Since the series

$$S(\theta) = 1 + \sum_{n=1}^{\infty} n^{-(\delta+1)} e^{in\theta}$$

is absolutely convergent for $\mathrm{Re}\,\delta > 0$, which we also concluded above by Raabe's test, and divergent if $\mathrm{Re}\,\delta \leq -1$, the hypergeometric series is absolutely convergent or divergent under the same conditions [22].

It remains to investigate what happens when $-1 < \mathrm{Re}\,\delta \leq 0$. We conclude that the hypergeometric series under this condition differs from the series $S(\theta)$ by an absolutely convergent series, so it suffices to prove the convergence of the sum $S(\theta)$. Summation by parts[10] implies

$$(S(\theta) - 1)\left(e^{i\theta} - 1\right) = \sum_{n=1}^{\infty} n^{-(\delta+1)}\left(e^{i(n+1)\theta} - e^{in\theta}\right)$$

$$= -e^{i\theta} - \sum_{n=1}^{\infty} e^{i(n+1)\theta}\left((n+1)^{-(\delta+1)} - n^{-(\delta+1)}\right) \tag{5.3}$$

The parenthesis in the last sum is bounded by

$$\left|(n+1)^{-(\delta+1)} - n^{-(\delta+1)}\right| = \left|(\delta+1)\int_n^{n+1} t^{-(\delta+2)}\,dt\right| \leq |\delta+1| n^{-(\mathrm{Re}\,\delta+2)}$$

---

[10] Summation by parts is

$$\sum_{k=m}^{n} f_k(g_{k+1} - g_k) = f_{n+1}g_{n+1} - f_m g_m - \sum_{k=m}^{n} g_{k+1}(f_{k+1} - f_k)$$

The last series on the right-hand side of (5.3) therefore converges absolutely and we conclude that $S(\theta)\left(e^{i\theta}-1\right)$ is finite, and $S(\theta)$ is convergent provided $\theta \neq 0, 2\pi$, i.e., $z \neq 1$.

We note that if α or β is a non-positive integer $-n$, the coefficient $a_k = 0, k > n$, and the hypergeometric series is a polynomial of degree n, and the proof is completed. $\square$

Comment 5.1. The results of Theorem 5.1 can also be obtained by employing Theorem B.4 and Corollary B.1 on pages 191 and 193, respectively. Even if no additional results are obtained by employing these theorems, it is illustrative to apply an alternative approach to determine the radius of convergence of the hypergeometric series.

We begin by identifying the recursion relation of the coefficient a_n. From (5.2), we obtain

$$a_{n+1} = \frac{(n+\alpha)(n+\beta)}{(n+1)(n+\gamma)} a_n = A_n a_n$$

where

$$\begin{cases} A_n = \dfrac{(n+\alpha)(n+\beta)}{(n+1)(n+\gamma)} = 1 + \dfrac{\alpha+\beta-\gamma-1}{n} + O(1/n^2) \\[2mm] = 1 - \dfrac{\delta+1}{n} + O(1/n^2) \\[2mm] B_n = 0 \end{cases}$$

which implies, using the notation in Theorem B.4,

$$\begin{cases} \alpha = 0 \\ \beta_1 = -(\delta+1) \\ \gamma_1 = 0 \end{cases}$$

and from the same theorem we obtain the dominant contribution of the sequence $\{a_n\}_{n=0}^{\infty}$ at large values of n, i.e.,

$$a_n = C n^{(\beta_1+\gamma_1)/(1-\alpha)} \left(1 + \frac{c}{n} + O(1/n^2)\right)$$

$$= C n^{-(\delta+1)} \left(1 + \frac{c}{n} + O(1/n^2)\right), \quad n \to \infty$$

From the results above and Corollary B.1, the asymptotic behavior becomes

$$\lim_{n\to\infty} \frac{a_{n+1}}{a_n} = 1$$

The convergence inside the unit circle is therefore established.

A necessary condition for convergence of the hypergeometric series on the unit circle can also be obtained from Corollary B.1. The result is

$$\frac{a_{n+1}}{a_n} = 1 - \frac{\delta+1}{n} + O(1/n^2)$$

By Raabe's test, see footnote 9, we get convergence on the unit circle provided $\mathrm{Re}\,\delta > 0$. ∎

We also conclude from the series representation of the hypergeometric function that $F(\alpha, \beta; \gamma; z)$ is analytic in α and β provided $|z| < 1$.

If γ is a positive integer,[11] the two roots of the indicial equation differ by an integer, and the general solution to the hypergeometric differential equation is obtained by the method developed in Section 2.4.3. The general solution then has the form

$$v(z) = G(\alpha, \beta, \gamma, z) + \ln z\, H(\alpha, \beta, \gamma, z)$$

where $G(\alpha, \beta, \gamma, z)$ and $H(\alpha, \beta, \gamma, z)$ are analytic in a neighborhood of $z = 0$.

We end this section by constructing alternative solutions to the hypergeometric differential equation. Formally, in the notion of Riemann's P symbol, the hypergeometric function belongs to the set

$$F(\alpha, \beta; \gamma; z) \in P \left\{ \begin{matrix} 0 & 1 & \infty \\ 0 & 0 & \alpha & z \\ 1-\gamma & \gamma-\alpha-\beta & \beta \end{matrix} \right\}$$

The displacement theorem, Theorem 3.3, in this notation reads, see (4.9) on page 55, with $\alpha_1 = \beta_2 = 0$

$$P \left\{ \begin{matrix} 0 & 1 & \infty \\ 0 & 0 & \alpha & z \\ 1-\gamma & \gamma-\alpha-\beta & \beta \end{matrix} \right\} = z^{1-\gamma} P \left\{ \begin{matrix} 0 & 1 & \infty \\ 0 & 0 & \alpha+1-\gamma & z \\ \gamma-1 & \gamma-\alpha-\beta & \beta+1-\gamma \end{matrix} \right\}$$

from which we conclude that

$$z^{1-\gamma} F(\alpha+1-\gamma, \beta+1-\gamma; 2-\gamma; z) \tag{5.4}$$

is also a solution to the hypergeometric differential equation,[12] and this solution, in general, represents the second solution to the hypergeometric differential equation in a power series expansion at $z = 0$. Indeed, by Theorem 5.1, the hypergeometric series corresponding to the right-hand side converges for $|z| < 1$, and on the unit circle the convergence is guaranteed, provided $\mathrm{Re}((2-\gamma) - (\alpha+1-\gamma) - (\beta+1-\gamma)) = \mathrm{Re}(\gamma-\alpha-\beta) > 0$.

Similarly,[13] using $\alpha_2 = \beta_1 = 0$ in (4.9),

$$P \left\{ \begin{matrix} 0 & 1 & \infty \\ 0 & 0 & \alpha & z \\ 1-\gamma & \gamma-\alpha-\beta & \beta \end{matrix} \right\} = (1-z)^{\gamma-\alpha-\beta} P \left\{ \begin{matrix} 0 & 1 & \infty \\ 0 & 0 & \gamma-\beta & z \\ 1-\gamma & \alpha+\beta-\gamma & \gamma-\alpha \end{matrix} \right\}$$

---

[11] See footnote 7 for the case $\gamma = 0$.

[12] Note that the corresponding hypergeometric series is well defined when $\gamma = 0, -1, -2, \ldots$, which was a case where the technique in Theorem 5.1 did not work, cf. also Lemma 4.1 on page 50.

[13] It does not matter if we write $(1-z)^{\gamma-\alpha-\beta}$ or $(z-1)^{\gamma-\alpha-\beta}$ since Riemann's P symbol denotes the set of solutions.

and, using $\alpha_1 = \alpha_2 = 0$ in (4.9),

$$P\left\{\begin{matrix} 0 & 1 & \infty \\ 0 & 0 & \alpha \\ 1-\gamma & \gamma-\alpha-\beta & \beta \end{matrix}\; z\right\}$$

$$= z^{1-\gamma}(1-z)^{\gamma-\alpha-\beta} P\left\{\begin{matrix} 0 & 1 & \infty \\ 0 & 0 & 1-\beta \\ \gamma-1 & \alpha+\beta-\gamma & 1-\alpha \end{matrix}\; z\right\} \tag{5.5}$$

which lead to

$$(1-z)^{\gamma-\alpha-\beta} F(\gamma-\alpha, \gamma-\beta; \gamma; z) \tag{5.6}$$

and

$$z^{1-\gamma}(1-z)^{\gamma-\alpha-\beta} F(1-\alpha, 1-\beta; 2-\gamma; z) \tag{5.7}$$

also are solutions to the same equation. The systematics of this observation is further developed in Section 5.4.

5.3 Recursion and differentiation formulae

To simplify the analysis in this section, we introduce the notion

$$F_\gamma(z) = F(\alpha, \beta; \gamma; z)$$

and form the difference $F_\gamma - F_{\gamma-1}$. By the use of the result in Theorem 5.1, we get

$$F_\gamma - F_{\gamma-1} = \sum_{n=0}^{\infty} \frac{(\alpha, n)(\beta, n)}{(1, n)(\gamma, n)} z^n \left(1 - \frac{\gamma+n-1}{\gamma-1}\right)$$

$$= -\frac{z}{\gamma-1} \sum_{n=0}^{\infty} \frac{(\alpha, n)(\beta, n)}{(1, n)(\gamma, n)} nz^{n-1} = -\frac{z}{\gamma-1} \frac{\mathrm{d}}{\mathrm{d}z} F(\alpha, \beta; \gamma; z)$$

We then have

$$zF_\gamma' = -(\gamma-1)(F_\gamma - F_{\gamma-1}) \tag{5.8}$$

It is convenient to also introduce the notation

$$a_n(\gamma) = \frac{(\alpha, n)(\beta, n)}{(1, n)(\gamma, n)}$$

Differentiation of the hypergeometric series gives

$$zF_\gamma' = \sum_{n=1}^{\infty} a_n(\gamma) nz^n = \sum_{n=0}^{\infty} a_{n+1}(\gamma)(n+1)z^{n+1} \tag{5.9}$$

We also observe that

$$\frac{a_{n+1}(\gamma)(n+1)}{a_n(\gamma)} = \frac{(n+\alpha)(n+\beta)}{n+\gamma} = n+\alpha+\beta-\gamma+\frac{(\alpha-\gamma)(\beta-\gamma)}{n+\gamma}$$

and

$$\frac{a_n(\gamma)}{n+\gamma} = \frac{(\alpha,n)(\beta,n)}{(1,n)(\gamma,n)(n+\gamma)} = \frac{(\alpha,n)(\beta,n)}{(1,n)(\gamma+1,n)\gamma} = \frac{a_n(\gamma+1)}{\gamma}$$

In addition, we also have (replace $\gamma \to \gamma-1$ in the last equality)

$$a_n(\gamma)(n+\gamma-1) = (\gamma-1)a_n(\gamma-1)$$

From these last three equalities, we can now conclude that

$$a_{n+1}(\gamma)(n+1) = a_n(\gamma)(n+\alpha+\beta-\gamma) + a_n(\gamma)\frac{(\alpha-\gamma)(\beta-\gamma)}{n+\gamma}$$

$$= a_n(\gamma)(n+\gamma-1) + a_n(\gamma)(\alpha+\beta-2\gamma+1) + a_n(\gamma+1)\frac{(\alpha-\gamma)(\beta-\gamma)}{\gamma}$$

$$= a_n(\gamma-1)(\gamma-1) + a_n(\gamma)(\alpha+\beta-2\gamma+1) + a_n(\gamma+1)\frac{(\alpha-\gamma)(\beta-\gamma)}{\gamma}$$

and we obtain from (5.9)

$$zF_\gamma' = z\left((\gamma-1)F_{\gamma-1} + (\alpha+\beta-2\gamma+1)F_\gamma + \frac{(\alpha-\gamma)(\beta-\gamma)}{\gamma}F_{\gamma+1}\right)$$

and from (5.8), we get

$$-(\gamma-1)(F_\gamma-F_{\gamma-1}) = z\left((\gamma-1)F_{\gamma-1}\right.$$
$$\left. + (\alpha+\beta-2\gamma+1)F_\gamma + \frac{(\alpha-\gamma)(\beta-\gamma)}{\gamma}F_{\gamma+1}\right)$$

or

$$(1-z)F_{\gamma-1} = \left(1+z\frac{\alpha+\beta-2\gamma+1}{\gamma-1}\right)F_\gamma + \frac{(\alpha-\gamma)(\beta-\gamma)}{\gamma(\gamma-1)}F_{\gamma+1}$$

The derivation of this recursion formula was done under the assumption that $|z| < 1$, so that all the power series in the derivation converge. However, the final result holds by analytic continuation for all $z \in \mathbb{C}$ in the common domain of analyticity of the left- and right-hand sides.

We summarize this result in a lemma.

Lemma 5.1. *The hypergeometric function satisfies*

$$(1-z)F(\alpha,\beta;\gamma-1;z) = \left(1+z\frac{\alpha+\beta-2\gamma+1}{\gamma-1}\right)F(\alpha,\beta;\gamma;z)$$
$$+ \frac{(\alpha-\gamma)(\beta-\gamma)}{\gamma(\gamma-1)}F(\alpha,\beta;\gamma+1;z)$$

in the common domain of analyticity of the left- and right-hand sides, and for all values of the parameters α, β, and γ for which the left- and right-hand sides are defined.

Example 5.1. We illustrate the use of the recursion formula in Lemma 5.1 by evaluating the value of $F_\gamma(1)$, when neither γ nor $\gamma - \alpha - \beta$ are zero or a negative integer. The recursion formula implies

$$\frac{\alpha + \beta - \gamma}{\gamma - 1} F_\gamma(1) = -\frac{(\alpha - \gamma)(\beta - \gamma)}{\gamma(\gamma - 1)} F_{\gamma+1}(1)$$

or

$$\frac{F_\gamma(1)}{F_{\gamma+1}(1)} = \frac{(\gamma - \alpha)(\gamma - \beta)}{\gamma(\gamma - \alpha - \beta)}$$

or more generally, by recursion

$$\frac{F_\gamma(1)}{F_{\gamma+m}(1)} = \frac{F_\gamma(1)}{F_{\gamma+1}(1)} \frac{F_{\gamma+1}(1)}{F_{\gamma+2}(1)} \cdot \ldots \cdot \frac{F_{\gamma+m-1}(1)}{F_{\gamma+m}(1)} = \frac{(\gamma - \alpha, m)(\gamma - \beta, m)}{(\gamma, m)(\gamma - \alpha - \beta, m)}$$

where $m \in \mathbb{N}$, and $(\alpha, m) = \Gamma(\alpha + m)/\Gamma(\alpha)$ is the Appell symbol, see Appendix A on page 173. Therefore

$$F_\gamma(1) = \frac{(\gamma - \alpha, m)(\gamma - \beta, m)}{(\gamma, m)(\gamma - \alpha - \beta, m)} F_{\gamma+m}(1)$$

which we write as

$$F_\gamma(1) = A B_m F_{\gamma+m}(1)$$

where we have introduced the notation

$$\begin{cases} A = \dfrac{\Gamma(\gamma)\Gamma(\gamma - \alpha - \beta)}{\Gamma(\gamma - \alpha)\Gamma(\gamma - \beta)} \\[2ex] B_m = \dfrac{\Gamma(\gamma - \alpha + m)\Gamma(\gamma - \beta + m)}{\Gamma(\gamma + m)\Gamma(\gamma - \alpha - \beta + m)} \end{cases}$$

and

$$F_{\gamma+m}(1) = \sum_{n=0}^{\infty} \frac{(\alpha, n)(\beta, n)}{(1, n)(\gamma + m, n)}$$

Note that the series $F_{\gamma+m}(1)$ always is absolutely convergent for sufficiently large m (use the same technique as in the derivation of Theorem 5.1), since $\mathrm{Re}(\gamma + m - \alpha - \beta) > 0$. Evaluate B_m and $F_{\gamma+m}(1)$ as $m \to \infty$. By Stirling's formula, see (A.6) on page 166, we get to leading order

$$\ln \Gamma(\alpha + m) = \ln \sqrt{2\pi} + \left(\alpha + m - \frac{1}{2}\right) \ln(\alpha + m) - \alpha - m + O(1/m)$$

We obtain the limits

$$B_m = F_{\gamma+m}(1) = 1, \text{ as } m \to \infty$$

and we conclude that

$$F(\alpha,\beta,\gamma;1) = F_\gamma(1) = A = \frac{\Gamma(\gamma)\Gamma(\gamma-\alpha-\beta)}{\Gamma(\gamma-\alpha)\Gamma(\gamma-\beta)} \tag{5.10}$$

This expression makes sense as long as the arguments in the numerator are not a negative integer or zero, i.e., $\gamma \neq 0, -1, -2, \ldots$ and $\gamma - \alpha - \beta \neq 0, -1, -2, \ldots$. This result is also proved below by the use of the integral representation of the hypergeometric function, see Corollary 5.1 on page 78. ∎

5.3.1 Gauss' contiguous relations

Above, in Lemma 5.1, we found a relation between three hypergeometric functions of different γ index. Another type of relation between the hypergeometric functions of different indices, α, β, and γ, is the Gauss[14] contiguous relations. There is a whole set of such relations, and we limit ourselves to show one of them. The other relations are proved with the same technique.

Lemma 5.2. *The hypergeometric function satisfies*

$$\gamma(1-z)F(\alpha,\beta;\gamma;z) - \gamma F(\alpha-1,\beta;\gamma;z) + z(\gamma-\beta)F(\alpha,\beta;\gamma+1;z) = 0$$

in the common domain of analyticity of the terms on the left-hand side, and for all values of the parameters α, β, and γ for which the terms on the left-hand side are defined.

Proof. The lemma is proved by the use of the result in Theorem 5.1. We rewrite the left-hand side of the lemma as a power series, and then prove that the coefficients of this power series are all zero, i.e.,

$$\gamma(1-z)F(\alpha,\beta;\gamma;z) - \gamma F(\alpha-1,\beta;\gamma;z) + z(\gamma-\beta)F(\alpha,\beta;\gamma+1;z) = \sum_{k=0}^{\infty} a_k z^k \tag{5.11}$$

and prove that all $a_k = 0$. Indeed, the coefficient in front of the lowest order term, $k = 0$, in (5.11) is

$$a_0 = \gamma - \gamma = 0$$

and the coefficient in front of the linear term, $k = 1$, is

$$a_1 = \gamma\frac{\alpha\beta}{\gamma} - \gamma - \gamma\frac{(\alpha-1)\beta}{\gamma} + (\gamma-\beta) = 0$$

The coefficient in front of the kth power of z in (5.11), $k = 2, 3, \ldots$, is

---

[14] Johann Carl Friedrich Gauss (1777–1855). German mathematician.

$$a_k = \gamma \frac{(\alpha,k)(\beta,k)}{(\gamma,k)k!} - \gamma \frac{(\alpha,k-1)(\beta,k-1)}{(\gamma,k-1)(k-1)!} - \gamma \frac{(\alpha-1,k)(\beta,k)}{(\gamma,k)k!}$$

$$+ (\gamma-\beta) \frac{(\alpha,k-1)(\beta,k-1)}{(\gamma+1,k-1)(k-1)!}$$

$$= \frac{((\alpha,k)-(\alpha-1,k))(\beta,k)-k(\gamma+k-1-\gamma+\beta)(\alpha,k-1)(\beta,k-1)}{(\gamma+1,k-1)k!}$$

where we used $(a,k) = a(a+1,k-1)$ and $(a,k)(a+k) = (a,k+1) = a(a+1,k)$, see (A.9) in Appendix A. Further simplifications give

$$a_k = \frac{((\alpha,k)-(\alpha-1,k))(\beta,k)-k(\beta+k-1)(\alpha,k-1)(\beta,k-1)}{(\gamma+1,k-1)k!}$$

$$= \frac{(\beta,k)\{(\alpha,k)-(\alpha-1,k)-k(\alpha,k-1)\}}{(\gamma+1,k-1)k!}$$

$$= \frac{(\beta,k)\{(\alpha,k)-(\alpha+k-1)(\alpha,k-1)\}}{(\gamma+1,k-1)k!} = 0$$

and the lemma is proved.   □

5.4 Kummer's solutions to hypergeometric differential equation

We have already proved that a Möbius transformation of the independent variable z in the Papperitz equation is still a solution of the original equation. In fact, from Theorem 4.1 (the substitution theorem) we have

$$P \left\{ \begin{matrix} a_1 & a_2 & a_3 \\ \alpha_1 & \alpha_2 & \alpha_3 \ z \\ \beta_1 & \beta_2 & \beta_3 \end{matrix} \right\} = P \left\{ \begin{matrix} t(a_1) & t(a_2) & t(a_3) \\ \alpha_1 & \alpha_2 & \alpha_3 \ t(z) \\ \beta_1 & \beta_2 & \beta_3 \end{matrix} \right\}$$

where

$$t(z) = \frac{az+b}{cz+d}, \qquad ad-bc \neq 0$$

which in particular applies to the hypergeometric differential equation. In this section, we are going to exploit the consequences of this theorem in a systematic way.

The effects of the displacement theorem, Theorem 3.3, have already been utilized at the end of Section 5.2. Four different solutions,[15] $u_1(z) = F(\alpha,\beta;\gamma;z)$, $u_2(z)$ given by (5.4), $u_{13}(z)$ given by (5.6), and $u_{14}(z)$ given by (5.7), were discovered. We are now ready to invoke the consequences of the substitution theorem discussed above. The observations above lead to the important theorem:

---

[15] The numbers of these solutions follow the traditional way of numbering, which is given in Table 5.2.

Theorem 5.2 (Kummer[16]). *All four ways of displacement of the roots of the indicial equation in*

$$
P\left\{
\begin{array}{ccc}
0 & 1 & \infty \\
0 & 0 & \alpha\ z \\
1-\gamma & \gamma-\alpha-\beta & \beta
\end{array}
\right\}
$$

viz., displacements leading to the following multiplicative factors

$$
\left\{1, z^{1-\gamma}, (1-z)^{\gamma-\alpha-\beta}, z^{1-\gamma}(1-z)^{\gamma-\alpha-\beta}\right\}
$$

and all six different permutations, obtained by combinations of Möbius transforms, of the singular points, 0, 1, and ∞, viz.,

$$
\{(0,1,\infty),(1,\infty,0),(\infty,0,1),(1,0,\infty),(0,\infty,1),(\infty,1,0)\}
$$

are solutions to the hypergeometric differential equation. In total there are 24 different combinations, which are listed in Table 5.2. The different domains of analyticity guaranteed by the convergence of the hypergeometric series, corresponding to different solutions in Table 5.2, are listed in Table 5.3.

An example illustrates the result of the theorem.

Example 5.2. Let $t(z) = 1 - z$. Then

$$
P\left\{
\begin{array}{ccc}
0 & 1 & \infty \\
0 & 0 & \alpha\ z \\
1-\gamma & \gamma-\alpha-\beta & \beta
\end{array}
\right\}
= P\left\{
\begin{array}{ccc}
0 & 1 & \infty \\
0 & 0 & \alpha\ 1-z \\
\gamma-\alpha-\beta & 1-\gamma & \beta
\end{array}
\right\}
$$

and

$$
F(\alpha,\beta;1-\gamma+\alpha+\beta;1-z)
$$

is a solution to the hypergeometric differential equation, cf. solution $u_5(z)$ in Table 5.2. Similarly, the solutions $u_6(z)$, $u_{17}(z)$, and $u_{18}(z)$ are found by the use of the displacement theorem, Theorem 3.3. ∎

Since a second order ODE only has two linearly independent solutions, any three of the solutions in Theorem 5.2 must be linearly dependent. Thus, between any three of these combinations there is a relation, i.e., $au_i(z) = bu_j(z) + cu_k(z)$, $i \neq j \neq k = 1,2,3,\ldots,24$, where the constants a, b, and c can be determined from explicit values of the hypergeometric function at two different points z. To find the constants a, b, and c can be rather cumbersome in many of the cases. The integral representation by Barnes[17] in Section 5.6 is often very useful in this context. We give an example of a relation of this kind by stating the combinations between $u_1(z)$, $u_5(z)$, and $u_6(z)$. This relation is proved in Section 5.6.1 by the use of Barnes' integral representation. An alternative proof is also given in Example 5.3.

---

[16] Ernst Eduard Kummer (1810–1893), German mathematician.
[17] Ernest William Barnes (1874–1953), English mathematician.

Table 5.2 The 24 different solutions of the hypergeometric differential equation due to Kummer.

	Solution	Sing.
$u_1(z)$	$F(\alpha,\beta;\gamma;z)$	$0,1,\infty$
$u_2(z)$	$z^{1-\gamma}F(1+\alpha-\gamma,1+\beta-\gamma;2-\gamma;z)$	$0,1,\infty$
$u_3(z)$	$(-z)^{-\alpha}F\left(\alpha,1+\alpha-\gamma;1+\alpha-\beta;\frac{1}{z}\right)$	$\infty,1,0$
$u_4(z)$	$(-z)^{-\beta}F\left(\beta,1+\beta-\gamma;1+\beta-\alpha;\frac{1}{z}\right)$	$\infty,1,0$
$u_5(z)$	$F(\alpha,\beta;1-\gamma+\alpha+\beta;1-z)$	$1,0,\infty$
$u_6(z)$	$(1-z)^{\gamma-\alpha-\beta}F(\gamma-\alpha,\gamma-\beta;1+\gamma-\alpha-\beta;1-z)$	$1,0,\infty$
$u_7(z)$	$(1-z)^{-\alpha}F\left(\alpha,\gamma-\beta;\gamma;\frac{z}{z-1}\right)$	$0,\infty,1$
$u_8(z)$	$z^{1-\gamma}(1-z)^{\gamma-\alpha-1}F\left(1+\alpha-\gamma,1-\beta;2-\gamma;\frac{z}{z-1}\right)$	$0,\infty,1$
$u_9(z)$	$(1-z)^{-\alpha}F\left(\alpha,\gamma-\beta;1+\alpha-\beta;\frac{1}{1-z}\right)$	$1,\infty,0$
$u_{10}(z)$	$(1-z)^{-\beta}F\left(\beta,\gamma-\alpha;1+\beta-\alpha;\frac{1}{1-z}\right)$	$1,\infty,0$
$u_{11}(z)$	$z^{-\alpha}F\left(\alpha,1+\alpha-\gamma;1+\alpha+\beta-\gamma;1-\frac{1}{z}\right)$	$\infty,0,1$
$u_{12}(z)$	$z^{\alpha-\gamma}(1-z)^{\gamma-\alpha-\beta}F\left(\gamma-\alpha,1-\alpha;1+\gamma-\alpha-\beta;1-\frac{1}{z}\right)$	$\infty,0,1$
$u_{13}(z)$	$(1-z)^{\gamma-\alpha-\beta}F(\gamma-\alpha,\gamma-\beta;\gamma;z)$	$0,1,\infty$
$u_{14}(z)$	$z^{1-\gamma}(1-z)^{\gamma-\alpha-\beta}F(1-\alpha,1-\beta;2-\gamma;z)$	$0,1,\infty$
$u_{15}(z)$	$z^{\beta-\gamma}(1-z)^{\gamma-\alpha-\beta}F\left(1-\beta,\gamma-\beta;1+\alpha-\beta;\frac{1}{z}\right)$	$\infty,1,0$
$u_{16}(z)$	$z^{\alpha-\gamma}(1-z)^{\gamma-\alpha-\beta}F\left(1-\alpha,\gamma-\alpha;1-\beta+\alpha;\frac{1}{z}\right)$	$\infty,1,0$
$u_{17}(z)$	$z^{1-\gamma}F(1+\alpha-\gamma,1+\beta-\gamma;1+\alpha+\beta-\gamma;1-z)$	$1,0,\infty$
$u_{18}(z)$	$z^{1-\gamma}(1-z)^{\gamma-\alpha-\beta}F(1-\alpha,1-\beta;1-\alpha-\beta+\gamma;1-z)$	$1,0,\infty$
$u_{19}(z)$	$(1-z)^{-\beta}F\left(\gamma-\alpha,\beta;\gamma;\frac{z}{z-1}\right)$	$0,\infty,1$
$u_{20}(z)$	$z^{1-\gamma}(1-z)^{\gamma-\beta-1}F\left(1+\beta-\gamma,1-\alpha;2-\gamma;\frac{z}{z-1}\right)$	$0,\infty,1$
$u_{21}(z)$	$z^{1-\gamma}(1-z)^{\gamma-\alpha-1}F\left(1-\beta,1+\alpha-\gamma;1+\alpha-\beta;\frac{1}{1-z}\right)$	$1,\infty,0$
$u_{22}(z)$	$z^{1-\gamma}(1-z)^{\gamma-\beta-1}F\left(1-\alpha,1+\beta-\gamma;1+\beta-\alpha;\frac{1}{1-z}\right)$	$1,\infty,0$
$u_{23}(z)$	$z^{-\beta}F\left(1+\beta-\gamma,\beta;1+\alpha+\beta-\gamma;1-\frac{1}{z}\right)$	$\infty,0,1$
$u_{24}(z)$	$z^{\beta-\gamma}(1-z)^{\gamma-\alpha-\beta}F\left(\gamma-\beta,1-\beta;1+\gamma-\alpha-\beta;1-\frac{1}{z}\right)$	$\infty,0,1$

Table 5.3 The domain of convergence of the hypergeometric series corresponding to different solutions in Table 5.2.

Solution #	Domain
1, 2, 13, 14	$|z| < 1$
3, 4, 15, 16	$|z| > 1$
5, 6, 17, 18	$|z-1| < 1$
7, 8, 19, 20	$\mathrm{Re}\,z < 1/2$
9, 10, 21, 22	$|z-1| > 1$
11, 12, 23, 24	$\mathrm{Re}\,z > 1/2$

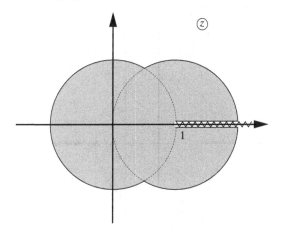

Fig. 5.1 The analytic continuity of the hypergeometric function $F(\alpha,\beta;\gamma;z)$ due to relation (5.12). The branch cut along the real axis vanishes if $\gamma - \alpha - \beta$ is a positive integer.

$$F(\alpha,\beta;\gamma;z) = \frac{\Gamma(\gamma)\Gamma(\gamma - \alpha - \beta)}{\Gamma(\gamma - \alpha)\Gamma(\gamma - \beta)} F(\alpha,\beta;1 - \gamma + \alpha + \beta; 1 - z)$$
$$+ \frac{\Gamma(\gamma)\Gamma(\alpha + \beta - \gamma)}{\Gamma(\alpha)\Gamma(\beta)} (1 - z)^{\gamma - \alpha - \beta} F(\gamma - \alpha, \gamma - \beta; 1 + \gamma - \alpha - \beta; 1 - z)$$

$$(5.12)$$

which extends the domain of analyticity of the hypergeometric function, see Figure 5.1.

Other relations, obtained from the solutions in Table 5.2, can be used to extend the domain of analyticity of the hypergeometric function further. For example, the result of Lemma 5.3 on page 79 is used to extend the domain of analyticity of the hypergeometric function as given in Figure 5.2, since

$$\left|\frac{z}{z - 1}\right| < 1 \quad \Leftrightarrow \quad |z - 1| > |z| \quad \Rightarrow \quad \mathrm{Re}\, z = x < \frac{1}{2}$$

Example 5.3. A proof of (5.12) is given in Section 5.6.1 by the use of Barnes' integral representation. In this example, we give an alternative proof of the relation by using (5.10), when neither γ nor $\gamma - \alpha - \beta$ are zero or a negative integer.

$$F(\alpha,\beta,\gamma;1) = \frac{\Gamma(\gamma)\Gamma(\gamma - \alpha - \beta)}{\Gamma(\gamma - \alpha)\Gamma(\gamma - \beta)}$$

and

$$F(\alpha,\beta,\gamma;0) = 1$$

We know that $u_1(z)$, $u_5(z)$, and $u_6(z)$ must be linearly dependent, i.e., there are constants a and b such that

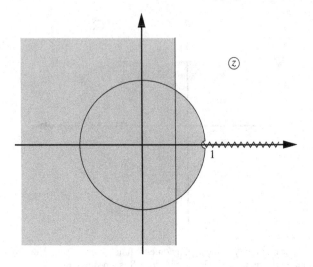

Fig. 5.2 The analytic continuity of the hypergeometric function $F(\alpha,\beta;\gamma;z)$ due to solution $u_7(z)$.

$$F(\alpha,\beta;\gamma;z) = aF(\alpha,\beta;1-\gamma+\alpha+\beta;1-z)$$
$$+b(1-z)^{\gamma-\alpha-\beta}F(\gamma-\alpha,\gamma-\beta;1+\gamma-\alpha-\beta;1-z)$$

Evaluate at $z = 0, 1$ and we get

$$1 = a\frac{\Gamma(1-\gamma+\alpha+\beta)\Gamma(1-\gamma)}{\Gamma(1-\gamma+\beta)\Gamma(1-\gamma+\alpha)} + b\frac{\Gamma(1+\gamma-\alpha-\beta)\Gamma(1-\gamma)}{\Gamma(1-\beta)\Gamma(1-\alpha)}$$

and

$$\frac{\Gamma(\gamma)\Gamma(\gamma-\alpha-\beta)}{\Gamma(\gamma-\alpha)\Gamma(\gamma-\beta)} = a$$

This implies using (A.4) on page 165

$$b\frac{\sin\pi\alpha\sin\pi\beta}{\sin\pi\gamma\sin\pi(\alpha+\beta-\gamma)}\frac{\Gamma(\beta)\Gamma(\alpha)}{\Gamma(\alpha+\beta-\gamma)\Gamma(\gamma)}$$
$$= 1 - \frac{\Gamma(\gamma)\Gamma(\gamma-\alpha-\beta)}{\Gamma(\gamma-\alpha)\Gamma(\gamma-\beta)}\frac{\Gamma(1-\gamma+\alpha+\beta)\Gamma(1-\gamma)}{\Gamma(1-\gamma+\beta)\Gamma(1-\gamma+\alpha)}$$

or with further use of (A.4)

$$b\frac{\sin\pi\alpha\sin\pi\beta}{\sin\pi\gamma\sin\pi(\alpha+\beta-\gamma)}\frac{\Gamma(\beta)\Gamma(\alpha)}{\Gamma(\alpha+\beta-\gamma)\Gamma(\gamma)}$$
$$= 1 - \frac{\sin\pi(\gamma-\alpha)\sin\pi(\gamma-\beta)}{\sin\pi\gamma\sin\pi(\gamma-\alpha-\beta)}$$
$$= \frac{\sin\pi\gamma\sin\pi(\gamma-\alpha-\beta) - \sin\pi(\gamma-\alpha)\sin\pi(\gamma-\beta)}{\sin\pi\gamma\sin\pi(\gamma-\alpha-\beta)}$$

which we simplify further

$$b\sin\pi\alpha\sin\pi\beta\,\frac{\Gamma(\beta)\Gamma(\alpha)}{\Gamma(\alpha+\beta-\gamma)\Gamma(\gamma)}$$
$$=\sin\pi(\gamma-\alpha)\sin\pi(\gamma-\beta)-\sin\pi\gamma\sin\pi(\gamma-\alpha-\beta)$$
$$=\frac{1}{2}\cos\pi(\beta-\alpha)-\frac{1}{2}\cos\pi(\alpha+\beta)=\sin\pi\alpha\sin\pi\beta$$

We thus conclude

$$\begin{cases} a=\dfrac{\Gamma(\gamma)\Gamma(\gamma-\alpha-\beta)}{\Gamma(\gamma-\alpha)\Gamma(\gamma-\beta)}\\[2mm] b=\dfrac{\Gamma(\alpha+\beta-\gamma)\Gamma(\gamma)}{\Gamma(\beta)\Gamma(\alpha)} \end{cases}$$

and we get

$$F(\alpha,\beta;\gamma;z)=\frac{\Gamma(\gamma)\Gamma(\gamma-\alpha-\beta)}{\Gamma(\gamma-\alpha)\Gamma(\gamma-\beta)}F(\alpha,\beta;1-\gamma+\alpha+\beta;1-z)$$
$$+\frac{\Gamma(\gamma)\Gamma(\alpha+\beta-\gamma)}{\Gamma(\alpha)\Gamma(\beta)}(1-z)^{\gamma-\alpha-\beta}F(\gamma-\alpha,\gamma-\beta;1+\gamma-\alpha-\beta;1-z)$$

■

5.5 Integral representation of $F(\alpha,\beta;\gamma;z)$

The hypergeometric function is conveniently expressed as the hypergeometric series in its domain of convergence, and the solutions in Table 5.2 provide a way of extending the domain of analyticity. Another means of extending the domain of analyticity is the use of integral representations. To this end, we prove an integral representation of the hypergeometric function.

Theorem 5.3. *The hypergeometric function has the integral representation*

$$F(\alpha,\beta;\gamma;z)=\frac{\Gamma(\gamma)}{\Gamma(\beta)\Gamma(\gamma-\beta)}\int_0^1 u^{\beta-1}(1-u)^{\gamma-\beta-1}(1-uz)^{-\alpha}\,du$$

valid for all $\operatorname{Re}\gamma>\operatorname{Re}\beta>0$ *and all* $\mathbb{C}\ni z\notin[1,\infty)$.

Proof. We prove that the right-hand side is identical to the power series expansion of the hypergeometric function, i.e., the hypergeometric series

$$\sum_{n=0}^{\infty}\frac{(\alpha,n)(\beta,n)}{(\gamma,n)}\frac{z^n}{n!}$$

For $|z| < 1$ and $u \in [0,1]$ expand the integrand,[18] see also Problem 5.4

$$(1-uz)^{-\alpha} = \sum_{n=0}^{\infty} \binom{-\alpha}{n}(-1)^n u^n z^n = \sum_{n=0}^{\infty}(\alpha,n)u^n\frac{z^n}{n!}$$

since the binomial coefficient is, see (A.14) on page 176

$$\binom{-\alpha}{n} = \frac{\Gamma(-\alpha+1)}{\Gamma(-\alpha+1-n)(1,n)} = (-1)^n\frac{(\alpha,n)}{n!}$$

where the last identity is obtained using (A.4) on page 165 twice, i.e.,

$$\frac{\Gamma(-\alpha+1)}{\Gamma(-\alpha+1-n)} = (-1)^n\frac{\Gamma(\alpha+n)}{\Gamma(\alpha)} = (-1)^n(\alpha,n)$$

We get

$$\int_0^1 u^{\beta-1}(1-u)^{\gamma-\beta-1}(1-uz)^{-\alpha}\,du = \sum_{n=0}^{\infty}(\alpha,n)\frac{z^n}{n!}\underbrace{\int_0^1 u^{\beta-1}(1-u)^{\gamma-\beta-1}u^n\,du}_{\frac{\Gamma(n+\beta)\Gamma(\gamma-\beta)}{\Gamma(\gamma+n)}}$$

provided $\mathrm{Re}(\gamma-\beta) > 0$ and $\mathrm{Re}\,\beta > 0$, since, see (A.16) and (A.17) on page 176,

$$\int_0^1 t^{x-1}(1-t)^{y-1}\,dt = \frac{\Gamma(x)\Gamma(y)}{\Gamma(x+y)}, \qquad \mathrm{Re}\,x, \mathrm{Re}\,y > 0$$

The right-hand side of the theorem then becomes

$$\frac{\Gamma(\gamma)}{\Gamma(\beta)\Gamma(\gamma-\beta)}\sum_{n=0}^{\infty}(\alpha,n)\frac{\Gamma(n+\beta)\Gamma(\gamma-\beta)}{\Gamma(\gamma+n)}\frac{z^n}{n!} = \sum_{n=0}^{\infty}\frac{(\alpha,n)(\beta,n)}{(\gamma,n)}\frac{z^n}{n!}$$

which is identical to the hypergeometric series. The integral thus coincides with the hypergeometric series inside the unit circle $|z| < 1$, and by analytic continuation to the larger region. □

Note that the indices α and β can be interchanged without affecting the result.

Corollary 5.1 (Gauss' formula). *For* $\gamma \neq 0, -1, -2, \ldots$ *and* $\mathrm{Re}(\gamma - \alpha - \beta) > 0$, *we have*

$$F(\alpha,\beta;\gamma;1) = \frac{\Gamma(\gamma)\Gamma(\gamma-\alpha-\beta)}{\Gamma(\gamma-\alpha)\Gamma(\gamma-\beta)}$$

---

[18] This identity can also be found by observing that $u(z) = (1-z)^{-\alpha}$ solves

$$z(z-1)u''(z) + [(\alpha+1)z]u'(z) = 0$$

which implies that

$$(1-z)^{-\alpha} = F(\alpha,0;0;z) = \sum_{n=0}^{\infty}(\alpha,n)\frac{z^n}{n!}$$

Proof. This identity is obtained from the integral representation in Theorem 5.3

$$F(\alpha,\beta;\gamma;1) = \frac{\Gamma(\gamma)}{\Gamma(\beta)\Gamma(\gamma-\beta)} \int_0^1 u^{\beta-1}(1-u)^{\gamma-\alpha-\beta-1}\,du$$

which is convergent when $\text{Re}\,\beta > 0$ and $\text{Re}(\gamma-\alpha-\beta) > 0$. The corollary is proved provided we can show

$$\int_0^1 u^{\beta-1}(1-u)^{\gamma-\alpha-\beta-1}\,du = \frac{\Gamma(\beta)\Gamma(\gamma-\alpha-\beta)}{\Gamma(\gamma-\alpha)}$$

But this latter integral is the beta function in Appendix A, see (A.16) and (A.17) on page 176. The condition on $\text{Re}\,\beta > 0$ can be relaxed, since the factor $\Gamma(\beta)$ cancels. □

A simple change of variable of integration $(u \to 1/u)$ proves the following corollary:

Corollary 5.2. *The hypergeometric series has the integral representation*

$$F(\alpha,\beta;\gamma;z) = \frac{\Gamma(\gamma)}{\Gamma(\beta)\Gamma(\gamma-\beta)} \int_1^\infty u^{\alpha-\gamma}(u-1)^{\gamma-\beta-1}(u-z)^{-\alpha}\,du$$

valid for all $\text{Re}\,\gamma > \text{Re}\,\beta > 0$ *and all* $\mathbb{C} \ni z \notin [1,\infty)$.

Lemma 5.3 (Euler[19]). *The hypergeometric function satisfies*

$$F(\alpha,\beta;\gamma;z) = (1-z)^{-\alpha}F\left(\alpha,\gamma-\beta;\gamma;\frac{z}{z-1}\right)$$

Proof. From the integral representation in Theorem 5.3, we have $(\text{Re}\,\gamma > \text{Re}\,\beta > 0$ and all $\mathbb{C} \ni z \notin [1,\infty))$

$$F(\alpha,\beta;\gamma;z) = \frac{\Gamma(\gamma)}{\Gamma(\beta)\Gamma(\gamma-\beta)} \int_0^1 u^{\beta-1}(1-u)^{\gamma-\beta-1}(1-uz)^{-\alpha}\,du$$

Make a change of variable $u = 1-t$. We get

$$F(\alpha,\beta;\gamma;z) = \frac{\Gamma(\gamma)}{\Gamma(\beta)\Gamma(\gamma-\beta)} \int_0^1 (1-t)^{\beta-1}t^{\gamma-\beta-1}(1-(1-t)z)^{-\alpha}\,dt$$

which we rewrite as

$$F(\alpha,\beta;\gamma;z)$$
$$= \frac{\Gamma(\gamma)}{\Gamma(\beta)\Gamma(\gamma-\beta)}(1-z)^{-\alpha} \int_0^1 t^{\gamma-\beta-1}(1-t)^{\gamma-(\gamma-\beta)-1}\left(1-t\frac{z}{z-1}\right)^{-\alpha}\,dt$$

Use the integral representation in Theorem 5.3 once again to conclude

---

[19] Leonhard Paul Euler (1707–1783), Swiss mathematician and physicist.

$$F(\alpha,\beta;\gamma;z) = \frac{\Gamma(\gamma)}{\Gamma(\beta)\Gamma(\gamma-\beta)}(1-z)^{-\alpha}\frac{\Gamma(\beta)\Gamma(\gamma-\beta)}{\Gamma(\gamma)}F\left(\alpha,\gamma-\beta;\gamma;\frac{z}{z-1}\right)$$

and analytic continuation of both sides in the equality proves the lemma. $\square$

The result in Lemma 5.3 then extends the domain analyticity of the hypergeometric series, see Figure 5.2. Moreover, the lemma expresses the linear relationship between solutions $u_1(z)$ and $u_7(z)$ in Table 5.2.

5.6 Barnes' integral representation

We are now ready to present an integral representation of the hypergeometric function due to Barnes.[20] More extensive literature on this subject is found in, e.g., Ref. 19.

Theorem 5.4 (Barnes). *If none of* α, β, *or* γ *are zero or a negative integer, and* $\alpha - \beta$ *is not an integer, the hypergeometric function can be represented as a contour integral*

$$F(\alpha,\beta;\gamma;z) = \frac{1}{2\pi i}\int_L \frac{\Gamma(\alpha+s)\Gamma(\beta+s)\Gamma(\gamma)}{\Gamma(\alpha)\Gamma(\beta)\Gamma(\gamma+s)}\Gamma(-s)(-z)^s\,ds, \quad |\arg(-z)| < \pi$$

where L *is a contour in the complex* s-*plane, starting at* $-i\infty$ *and ending at* $i\infty$, *such that all poles to* $\Gamma(\alpha+s)\Gamma(\beta+s)$ *(i.e.,* $s = -\alpha, -\alpha-1, -\alpha-2,\dots$ *and* $s = -\beta, -\beta-1, -\beta-2,\dots$*) lie to the left and all poles to* $\Gamma(-s)$ *(i.e.,* $s \in \mathbb{N}$*) lie to the right of* L, *see Figure 5.3.*

Proof. The proof of this theorem is rather complex, and we prefer to break it down into two lemmas. The first one, Lemma 5.4, proves that the integral converges. The second one, Lemma 5.5, proves that the right-hand side coincides with the hypergeometric series in the domain $|z| < 1$, $|\arg(-z)| < \pi - \delta$, $\delta > 0$, and, therefore, by analytic continuation, is identical to the hypergeometric function in $|\arg(-z)| < \pi$. $\square$

Lemma 5.4. *The integral in Theorem 5.4 converges for all* z *such that* $|\arg(-z)| < \pi - \delta$, $\delta > 0$, *see Figure 5.4.*

Proof. We rewrite the gamma functions in the integrand using (A.4).

$$\frac{\Gamma(\alpha+s)\Gamma(\beta+s)}{\Gamma(\gamma+s)}\Gamma(-s) = -\frac{\Gamma(\alpha+s)}{\Gamma(s)}\frac{\Gamma(\beta+s)}{\Gamma(s)}\frac{\Gamma(s)}{\Gamma(\gamma+s)}\frac{\Gamma(s)}{\Gamma(1+s)}\frac{\pi}{\sin\pi s}$$

---

[20] The integral is also called a Mellin–Barnes' integral. Robert Hjalmar Mellin (1854–1933) was a Finnish mathematician, and Barnes was introduced in footnote 17 on page 73.

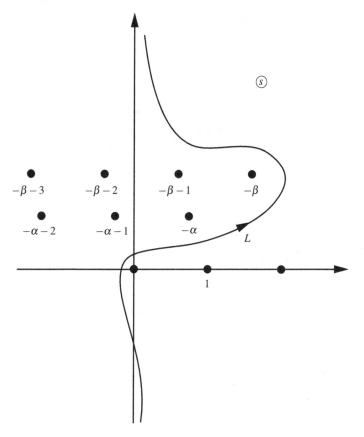

Fig. 5.3 The contour L in the complex s-plane used in Theorem 5.4.

Lemma A.4 on page 172 and Lemma A.1 on page 166 show that for each $\varepsilon \in (0, 1/2)$ we can estimate the gamma functions in the integrand as

$$\left| \frac{\Gamma(\alpha+s)\Gamma(\beta+s)}{\Gamma(\gamma+s)} \Gamma(-s) \right| \leq C e^{\mathrm{Re}(\alpha+\beta-\gamma-1)\ln|s|-\pi|\mathrm{Im}\,s|} \qquad (5.13)$$

when $s \in \{s \in \mathbb{C} : |s+\alpha+n| \geq \varepsilon, |s+\beta+n| \geq \varepsilon, |s-n| \geq \varepsilon, \forall n \in \mathbb{N}\}$. Note that (5.13) has no singularity at $s = -\gamma - n$ or $s = -n, n \in \mathbb{N}$.

We also have the following estimate:

$$\left| (-z)^s \right| = \left| |z|^s e^{\mathrm{i}s\arg(-z)} \right| = e^{\mathrm{Re}\,s\ln|z|-\mathrm{Im}\,s\arg(-z)}$$

The integrand can now be estimated. We get

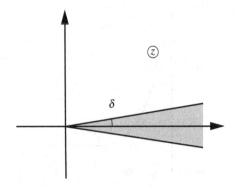

Fig. 5.4 The domain of convergence (unshaded region) of the integral in Theorem 5.4.

$$\left| \frac{\Gamma(\alpha+s)\Gamma(\beta+s)}{\Gamma(\gamma+s)}\Gamma(-s)(-z)^s \right| \leq Ce^{\mathrm{Re}(\alpha+\beta-\gamma-1)\ln|s|-\pi|\mathrm{Im}\,s|-\mathrm{Im}\,s\arg(-z)+\mathrm{Re}\,s\ln|z|}$$

$$(5.14)$$

when $s \in \{s \in \mathbb{C} : |s+\alpha+n| \geq \varepsilon, |s+\beta+n| \geq \varepsilon, |s-n| \geq \varepsilon, \forall n \in \mathbb{N}\}$. We conclude that the integrand vanishes by an exponential factor, viz.,

$$e^{(|\arg(-z)|-\pi)|\mathrm{Im}\,s|} \leq e^{-\delta|\mathrm{Im}\,s|}$$

as $s \to \pm i\infty$, provided $|\arg(-z)| < \pi - \delta$, $\delta > 0$, and the lemma is proved.   □

Lemma 5.5. *The integral in Theorem 5.4 is identical to the hypergeometric series in $|z| < 1$.*

Proof. We close the contour L by the semi-circle C in the right-hand side of the complex s-plane, see Figure 5.5. The poles at $s \in \mathbb{N}$ are avoided by parameterizing the contour C as[21]

$$C: s = (N+1/2)\,e^{i\phi} = (N+1/2)(\cos\phi + i\sin\phi), \qquad \phi \in [-\pi/2, \pi/2]$$

and let $N \to \infty$ through the integer numbers. As in the proof of Lemma 5.4, the integrand on the contour C can be estimated for large N, see (5.14),

$$\left| \frac{\Gamma(\alpha+s)\Gamma(\beta+s)}{\Gamma(\gamma+s)}\Gamma(-s)(-z)^s \right|$$
$$\leq Ce^{\mathrm{Re}(\alpha+\beta-\gamma-1)\ln(N+1/2)+(N+1/2)\{\ln|z|\cos\phi-\delta|\sin\phi|\}}$$

We have here assumed that $|\arg(-z)| < \pi - \delta$, $\delta > 0$. Notice that on the contour C, the variable $s \in \{s \in \mathbb{C} : |s+\alpha+n| \geq \varepsilon, |s+\beta+n| \geq \varepsilon, |s-n| \geq \varepsilon, \forall n \in \mathbb{N}\}$. For an $|z| < 1$, the dominant term in the exponent is

---

[21] To be precise, it is only in the limit, $N \to \infty$, that the contour C is a complete semi-circle, as the contour L in this limit approaches the imaginary axis.

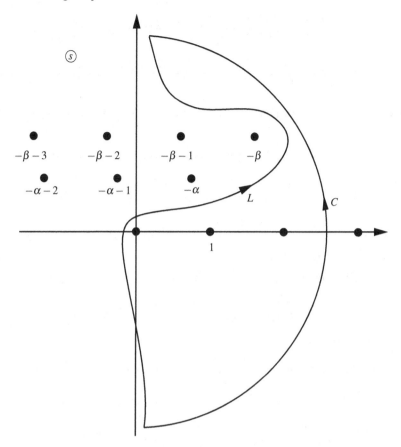

Fig. 5.5 The contours L and C in the complex s-plane.

$$(N + 1/2) \{\ln |z| \cos \phi - \delta |\sin \phi|\}$$

$$\leq (N + 1/2) \begin{cases} -\delta |\sin \phi| \leq -\delta/\sqrt{2} < 0, & |\phi| \in [\pi/4, \pi/2] \\ \ln |z| \cos \phi \leq \ln |z|/\sqrt{2} < 0, & |\phi| \in [0, \pi/4] \end{cases}$$

The contribution from the contour C therefore vanishes by an exponential factor as $N \to \infty$, and we have

$$\int_L \frac{\Gamma(\alpha + s)\Gamma(\beta + s)\Gamma(\gamma)}{\Gamma(\alpha)\Gamma(\beta)\Gamma(\gamma + s)} \Gamma(-s)(-z)^s \, ds$$

$$= \int_{L-C} \frac{\Gamma(\alpha + s)\Gamma(\beta + s)\Gamma(\gamma)}{\Gamma(\alpha)\Gamma(\beta)\Gamma(\gamma + s)} \Gamma(-s)(-z)^s \, ds$$

and consequently by the use of the residue theorem

$$\int_L \frac{\Gamma(\alpha+s)\Gamma(\beta+s)\Gamma(\gamma)}{\Gamma(\alpha)\Gamma(\beta)\Gamma(\gamma+s)}\Gamma(-s)(-z)^s \, ds$$

$$= -2\pi i \sum_{n=0}^{\infty} \operatorname*{Res}_{s=n} \frac{\Gamma(\alpha+s)\Gamma(\beta+s)\Gamma(\gamma)}{\Gamma(\alpha)\Gamma(\beta)\Gamma(\gamma+s)}\Gamma(-s)(-z)^s$$

The residues of $\Gamma(z)$ at the poles $z = 0, -1, -2, \ldots$ are, see (A.3) on page 164,

$$\operatorname*{Res}_{z=-n}\Gamma(z) = \frac{(-1)^n}{n!}, \quad n \in \mathbb{N}$$

The residues of $\Gamma(-s)$ then are[22]

$$\operatorname*{Res}_{s=n}\Gamma(-s) = -\frac{(-1)^n}{n!} \tag{5.15}$$

$$\int_L \frac{\Gamma(\alpha+s)\Gamma(\beta+s)\Gamma(\gamma)}{\Gamma(\alpha)\Gamma(\beta)\Gamma(\gamma+s)}\Gamma(-s)(-z)^s \, ds$$

$$= 2\pi i \sum_{n=0}^{\infty} \frac{\Gamma(\alpha+n)\Gamma(\beta+n)\Gamma(\gamma)}{\Gamma(\alpha)\Gamma(\beta)\Gamma(\gamma+n)}\frac{(-1)^n}{n!}(-z)^n = 2\pi i \sum_{n=0}^{\infty} \frac{(\alpha,n)(\beta,n)}{(\gamma,n)}\frac{z^n}{n!}$$

and the lemma is proved. $\square$

As a corollary to Barnes' theorem, Theorem 5.4, we get

Corollary 5.3.

$$\Gamma(\alpha)(1-z)^{-\alpha} = \frac{1}{2\pi i}\int_L \Gamma(\alpha+s)\Gamma(-s)(-z)^s \, ds, \quad |\arg(-z)| < \pi$$

Proof. Let $\beta = \gamma$ in Theorem 5.4. We get for $|\arg(-z)| < \pi$

$$F(\alpha,\beta;\beta;z) = \frac{1}{2\pi i}\int_L \frac{\Gamma(\alpha+s)}{\Gamma(\alpha)}\Gamma(-s)(-z)^s \, ds$$

The corollary then follows directly by the use of the result of Problem 5.4. $\square$

Barnes' theorem, Theorem 5.4, can be used to study the behavior of the solution $F(\alpha,\beta;\gamma;z)$ outside the unit circle $|z| < 1$. The first step is the following theorem, which also proves the linear relationship between solutions $u_1(z)$, $u_3(z)$ and $u_4(z)$ in Table 5.2:

Theorem 5.5. *For* $|\arg(-z)| < \pi$, *and* α, β, *or* γ *not zero or a negative integer, and* $\alpha - \beta$ *not an integer, the hypergeometric functions satisfy*

---

[22] Note that, if the residue of the meromorphic function $f(z)$ at $z = c$ is a_1 $(\lim_{z \to c}(z-c)f(z) = a_1)$, then the residue of $f(-s)$ at $s = -c$ is $-a_1$, since $\lim_{s \to -c}(s+c)f(-s) = -a_1$.

$$F(\alpha,\beta;\gamma;z) = \frac{\Gamma(\gamma)\Gamma(\beta-\alpha)}{\Gamma(\beta)\Gamma(\gamma-\alpha)}(-z)^{-\alpha} F\left(\alpha,1-\gamma+\alpha;1-\beta+\alpha;\frac{1}{z}\right)$$

$$+ \frac{\Gamma(\gamma)\Gamma(\alpha-\beta)}{\Gamma(\alpha)\Gamma(\gamma-\beta)}(-z)^{-\beta} F\left(\beta,1-\gamma+\beta;1-\alpha+\beta;\frac{1}{z}\right)$$

Proof. Apply Barnes' integral representation, and close the contour L in Figure 5.3 with a semi-circle C' to the **left** in the complex s-plane. For technical reasons, we divide the semi-circle into two parts, since our estimate of the quotients of the gamma functions (5.13) holds only for $|\arg(s)| \leq \pi$. Therefore

$$C': s = Re^{i\phi} = R(\cos\phi + i\sin\phi), \qquad \phi \in [\pi/2,\pi] \cup [-\pi,-\pi/2]$$

The closed contour then encloses the simple poles at $s = -\alpha - n$ and $s = -\beta - m$, where $m,n \in \mathbb{N}$, as $R \to \infty$. Use (5.13) to estimate the integrand, i.e.,

$$\left| \frac{\Gamma(\alpha+s)\Gamma(\beta+s)}{\Gamma(\gamma+s)}\Gamma(-s) \right| \leq Ce^{\text{Re}(\alpha+\beta-\gamma-1)\ln|s|-\pi|\text{Im}\,s|}$$

when $s \in \{s \in \mathbb{C}: |s+\alpha+n| \geq \varepsilon, |s+\beta+n| \geq \varepsilon, |s-n| \geq \varepsilon, \forall n \in \mathbb{N}\}$. We also have the following estimate, provided $|\arg(-z)| < \pi - \delta, \delta > 0$

$$\left|(-z)^s\right| = \left||z|^s e^{is\arg(-z)}\right| = e^{\text{Re}\,s\ln|z|-\text{Im}\,s\arg(-z)} \leq e^{\text{Re}\,s\ln|z|+|\text{Im}\,s|(\pi-\delta)}$$

and, as in the proof of Lemma 5.5, we estimate the integrand on the contour C' (for values of R that avoid the poles)

$$\left| \frac{\Gamma(\alpha+s)\Gamma(\beta+s)}{\Gamma(\gamma+s)}\Gamma(-s)(-z)^s \right| \leq Ce^{\text{Re}(\alpha+\beta-\gamma-1)\ln R+R\{\ln|z|\cos\phi-\delta|\sin\phi|\}}$$

For an $|z| > 1$, the dominant term in the exponent is

$$R\left(\ln|z|\cos\phi - \delta|\sin\phi|\right)$$

$$\leq R \begin{cases} -\delta|\sin\phi| \leq -\delta/\sqrt{2} < 0, & |\phi| \in [\pi/2,3\pi/4] \\ \ln|z|\cos\phi \leq -\ln|z|/\sqrt{2} < 0, & |\phi| \in [3\pi/4,\pi] \end{cases}$$

The contribution from the contour C' vanishes by an exponential factor as $R \to \infty$, and the residue theorem implies

$$\int_L \frac{\Gamma(\alpha+s)\Gamma(\beta+s)\Gamma(\gamma)}{\Gamma(\alpha)\Gamma(\beta)\Gamma(\gamma+s)}\Gamma(-s)(-z)^s \, ds$$

$$= \int_{L+C'} \frac{\Gamma(\alpha+s)\Gamma(\beta+s)\Gamma(\gamma)}{\Gamma(\alpha)\Gamma(\beta)\Gamma(\gamma+s)}\Gamma(-s)(-z)^s \, ds$$

$$= 2\pi i \sum_{n=0}^{\infty} \operatorname*{Res}_{\substack{\alpha+s=0,-1,\ldots \\ \beta+s=0,-1,\ldots}} \frac{\Gamma(\alpha+s)\Gamma(\beta+s)\Gamma(\gamma)}{\Gamma(\alpha)\Gamma(\beta)\Gamma(\gamma+s)}\Gamma(-s)(-z)^s$$

The residues of $\Gamma(z)$ at the poles $z = 0, -1, -2, \ldots$ are, see (A.3) on page 164,

$$\operatorname*{Res}_{z=-n} \Gamma(z) = \frac{(-1)^n}{n!}, \quad n \in \mathbb{N}$$

We get

$$\int_L \frac{\Gamma(\alpha+s)\Gamma(\beta+s)\Gamma(\gamma)}{\Gamma(\alpha)\Gamma(\beta)\Gamma(\gamma+s)}\Gamma(-s)(-z)^s \, ds$$

$$= 2\pi i \sum_{n=0}^{\infty} \frac{\Gamma(\beta-\alpha-n)\Gamma(\gamma)\Gamma(\alpha+n)}{\Gamma(\alpha)\Gamma(\beta)\Gamma(\gamma-\alpha-n)} \frac{(-1)^n}{n!}(-z)^{-\alpha-n} + (\alpha \leftrightarrow \beta)$$

$$= 2\pi i (-z)^{-\alpha} \frac{\Gamma(\beta-\alpha)\Gamma(\gamma)}{\Gamma(\gamma-\alpha)\Gamma(\beta)} \sum_{n=0}^{\infty} \frac{(1-\gamma+\alpha,n)(\alpha,n)}{(1-\beta+\alpha,n)} \frac{z^{-n}}{n!} + (\alpha \leftrightarrow \beta)$$

$$= 2\pi i (-z)^{-\alpha} \frac{\Gamma(\beta-\alpha)\Gamma(\gamma)}{\Gamma(\gamma-\alpha)\Gamma(\beta)} F\left(\alpha, 1-\gamma+\alpha; 1-\beta+\alpha; \frac{1}{z}\right) + (\alpha \leftrightarrow \beta)$$

since by (A.11) we have

$$\Gamma(\beta-\alpha-n) = \frac{(-1)^n \Gamma(\beta-\alpha)}{(1-\beta+\alpha,n)}$$

and similarly for $\Gamma(\gamma-\alpha-n)$. The theorem is proved for $|z| > 1$, but by analytic continuation, the result is valid for all values of $z \in \mathbb{C}$ that are in the common domain of analyticity of both the left- and the right-hand side of the identity. This ends the proof of the theorem. $\square$

Also very useful is the following lemma by Barnes:

Lemma 5.6 (Barnes' lemma).

$$\frac{1}{2\pi i}\int_B \Gamma(\alpha+s)\Gamma(\beta+s)\Gamma(\gamma-s)\Gamma(\delta-s)\,ds$$

$$= \frac{\Gamma(\alpha+\gamma)\Gamma(\alpha+\delta)\Gamma(\beta+\gamma)\Gamma(\beta+\delta)}{\Gamma(\alpha+\beta+\gamma+\delta)}$$

where B is a contour in the complex s-plane, starting at $-i\infty$ and ending at $i\infty$, such that all poles $s = -\alpha, -\alpha-1, -\alpha-2, \ldots$ and $s = -\beta, -\beta-1, -\beta-2, \ldots$ lie to the left and all poles $s = \gamma, \gamma+1, \gamma+2, \ldots$ and $s = \delta, \delta+1, \delta+2, \ldots$ lie to the right of B, see Figure 5.6. The poles of $\Gamma(\gamma-s)\Gamma(\delta-s)$ are all assumed simple, and they are assumed not to coincide with the poles of $\Gamma(\alpha+s)\Gamma(\beta+s)$.

Proof. We rewrite the gamma functions in the integrand using (A.4)

$$\Gamma(\alpha+s)\Gamma(\beta+s)\Gamma(\gamma-s)\Gamma(\delta-s) =$$

$$\frac{\Gamma(\alpha+s)}{\Gamma(s)}\frac{\Gamma(\beta+s)}{\Gamma(s)}\frac{\Gamma(s)}{\Gamma(1-\gamma+s)}\frac{\Gamma(s)}{\Gamma(1-\delta+s)}\frac{\pi^2}{\sin\pi(s-\gamma)\sin\pi(s-\delta)}$$

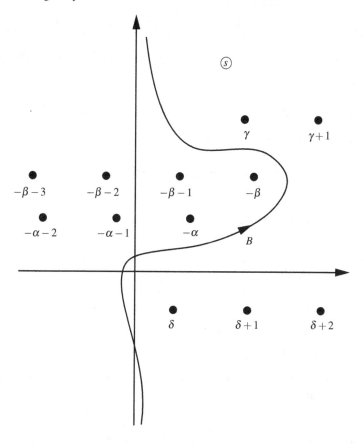

Fig. 5.6 The contour B in the complex s-plane used in Lemma 5.6.

Use Lemma A.4 on page 172 and Corollary A.1 on page 168 to estimate the integrand

$$|\Gamma(\alpha+s)\Gamma(\beta+s)\Gamma(\gamma-s)\Gamma(\delta-s)| \le Ce^{\mathrm{Re}(\alpha+\beta+\gamma+\delta-2)\ln|s|-2\pi|\mathrm{Im}\,s|}$$

for $s \in \{s \in \mathbb{C}: |s+\alpha+n| \ge \varepsilon, |s+\beta+n| \ge \varepsilon, |s-\gamma-n| \ge \varepsilon, |s-\delta-n| \ge \varepsilon, \forall n \in \mathbb{N}\}$. Note that this factor has no singularity at $s = -n$, $n \in \mathbb{N}$.

We first assume that $\mathrm{Re}\,(\alpha+\beta+\gamma+\delta-1) < 0$. The integrand on a semi-circle C, with radius R, in the right half plane of the complex s-plane, i.e.,

$$C: s = Re^{\mathrm{i}\phi} = R(\cos\phi + \mathrm{i}\sin\phi), \qquad \phi \in [-\pi/2, \pi/2]$$

that avoids the poles at $s = \gamma+n$ and $\delta+n$, has the leading contribution

$$|\Gamma(\alpha+s)\Gamma(\beta+s)\Gamma(\gamma-s)\Gamma(\delta-s)| \le Ce^{\mathrm{Re}(\alpha+\beta+\gamma+\delta-2)\ln R-2\pi R|\sin\phi|}$$

From this estimate, we conclude that the contribution from C vanishes at least as R^{-1} as $R \to \infty$, if $\mathrm{Re}(\alpha + \beta + \gamma + \delta - 1) < 0$.

The residues of the integrand at the poles at $s = \gamma + n$ and $\delta + n$ are computed by the use of (5.15)

$$\operatorname*{Res}_{s=\gamma+n} \Gamma(\gamma - s) = \operatorname*{Res}_{s=n} \Gamma(-s) = -\frac{(-1)^n}{n!}, \quad n \in \mathbb{N}$$

The residue theorem then gives

$$I = \frac{1}{2\pi i} \int_B \Gamma(\alpha + s)\Gamma(\beta + s)\Gamma(\gamma - s)\Gamma(\delta - s)\,ds$$

$$= \sum_{n=0}^{\infty} \frac{(-1)^n}{n!} \Gamma(\alpha + \gamma + n)\Gamma(\beta + \gamma + n)\Gamma(\delta - \gamma - n)$$

$$+ \sum_{n=0}^{\infty} \frac{(-1)^n}{n!} \Gamma(\alpha + \delta + n)\Gamma(\beta + \delta + n)\Gamma(\gamma - \delta - n)$$

$$= \sum_{n=0}^{\infty} \frac{(-1)^n}{n!} \frac{\Gamma(\alpha + \gamma + n)\Gamma(\beta + \gamma + n)}{\Gamma(1 + \gamma - \delta + n)} \frac{\pi}{\sin \pi(\delta - \gamma + n)}$$

$$+ \sum_{n=0}^{\infty} \frac{(-1)^n}{n!} \frac{\Gamma(\alpha + \delta + n)\Gamma(\beta + \delta + n)}{\Gamma(1 + \delta - \gamma + n)} \frac{\pi}{\sin \pi(\gamma - \delta + n)}$$

where we also have used (A.4). Gauss' formula in Corollary 5.1 on page 78 is then applied. We obtain

$$I = \frac{\pi}{\sin \pi(\delta - \gamma)} \sum_{n=0}^{\infty} \frac{\Gamma(\alpha + \gamma + n)\Gamma(\beta + \gamma + n)}{\Gamma(1 + \gamma - \delta + n)\,n!} + \sum_{n=0}^{\infty} (\gamma \leftrightarrow \delta)$$

$$= \frac{\pi}{\sin \pi(\delta - \gamma)} \frac{\Gamma(\alpha + \gamma)\Gamma(\beta + \gamma)}{\Gamma(1 + \gamma - \delta)} F(\alpha + \gamma, \beta + \gamma; 1 + \gamma - \delta; 1) + \sum_{n=0}^{\infty} (\gamma \leftrightarrow \delta)$$

$$= \frac{\pi}{\sin \pi(\delta - \gamma)} \frac{\Gamma(\alpha + \gamma)\Gamma(\beta + \gamma)\Gamma(1 - \alpha - \beta - \gamma - \delta)}{\Gamma(1 - \alpha - \delta)\Gamma(1 - \beta - \delta)} + \sum_{n=0}^{\infty} (\gamma \leftrightarrow \delta)$$

We rewrite using (A.4)

$$I = \frac{\pi\Gamma(1 - \alpha - \beta - \gamma - \delta)}{\sin \pi(\gamma - \delta)} \left(\frac{\Gamma(\alpha + \delta)\Gamma(\beta + \delta)}{\Gamma(1 - \alpha - \gamma)\Gamma(1 - \beta - \gamma)} \right.$$

$$\left. - \frac{\Gamma(\alpha + \gamma)\Gamma(\beta + \gamma)}{\Gamma(1 - \alpha - \delta)\Gamma(1 - \beta - \delta)} \right)$$

$$= \frac{\Gamma(\alpha + \gamma)\Gamma(\beta + \gamma)\Gamma(\alpha + \delta)\Gamma(\beta + \delta)}{\sin \pi(\gamma - \delta)\sin \pi(\alpha + \beta + \gamma + \delta)\Gamma(\alpha + \beta + \gamma + \delta)}$$

$$\times \{\sin \pi(\alpha + \gamma)\sin \pi(\beta + \gamma) - \sin \pi(\alpha + \delta)\sin \pi(\beta + \delta)\}$$

Since[23]

$$\sin \pi(\alpha + \beta + \gamma + \delta) \sin \pi(\gamma - \delta)$$
$$= \sin \pi(\alpha + \gamma) \sin \pi(\beta + \gamma) - \sin \pi(\alpha + \delta) \sin \pi(\beta + \delta)$$

we have finally proved

$$\frac{1}{2\pi i} \int_B \Gamma(\alpha + s)\Gamma(\beta + s)\Gamma(\gamma - s)\Gamma(\delta - s)\,ds$$
$$= \frac{\Gamma(\alpha + \gamma)\Gamma(\alpha + \delta)\Gamma(\beta + \gamma)\Gamma(\beta + \delta)}{\Gamma(\alpha + \beta + \gamma + \delta)}$$

for the parameters α, β, γ, and δ satisfying $\mathrm{Re}\,(\alpha + \beta + \gamma + \delta - 1) < 0$. However, by analytic continuation of the left- and right-hand sides in the parameters α, β, γ, and δ, the lemma is true for all values α, β, γ, and δ, in their common domain of analyticity. $\square$

5.6.1 Relation between $F(\cdot,\cdot;\cdot;z)$ and $F(\cdot,\cdot;\cdot;1-z)$

The proof of the relation between $u_1(z)$, $u_5(z)$, and $u_6(z)$ in Table 5.2, see (5.12), was postponed, and it is the purpose of this section to present a proof of this relation.

We start by Barnes' integral representation in Theorem 5.4, i.e.,

$$\frac{\Gamma(\alpha)\Gamma(\beta)}{\Gamma(\gamma)} F(\alpha, \beta; \gamma; z) = \frac{1}{2\pi i} \int_L \frac{\Gamma(\alpha + s)\Gamma(\beta + s)}{\Gamma(\gamma + s)} \Gamma(-s)(-z)^s\,ds$$

where $|\arg(-z)| < \pi$, and where the contour L is depicted in Figure 5.3. Moreover, use Barnes' lemma, Lemma 5.6, with $\gamma = s$ and $\delta = \gamma - \alpha - \beta$. We get

$$\frac{1}{2\pi i} \int_B \Gamma(\alpha + t)\Gamma(\beta + t)\Gamma(s - t)\Gamma(\gamma - \alpha - \beta - t)\,dt$$
$$= \frac{\Gamma(\alpha + s)\Gamma(\gamma - \beta)\Gamma(\beta + s)\Gamma(\gamma - \alpha)}{\Gamma(s + \gamma)}$$

---

[23] This identity can be derived by integration

$$\sin(a+c)\sin(b+c) - \sin(a+d)\sin(b+d) = -\int_c^d \frac{d}{dx}\sin(a+x)\sin(b+x)\,dx$$
$$= \frac{1}{2} \int_c^d \frac{d}{dx}(\cos(a+b+2x) - \cos(a-b))\,dx = \frac{1}{2}\cos(a+b+2x)\big|_{x=c}^{x=d}$$
$$= \frac{1}{2}\cos(a+b+2d) - \frac{1}{2}\cos(a+b+2c) = \sin(a+b+c+d)\sin(c-d)$$

where B is similar to the contour in Figure 5.6 (remember replacing $\gamma = s$ and $\delta = \gamma - \alpha - \beta$). We get

$$2\pi i \frac{\Gamma(\alpha)\Gamma(\beta)}{\Gamma(\gamma)} F(\alpha, \beta; \gamma; z)$$

$$= \int_L \left\{ \frac{1}{2\pi i} \int_B \Gamma(\alpha+t)\Gamma(\beta+t)\Gamma(s-t)\Gamma(\gamma-\alpha-\beta-t) \, dt \right\} \frac{\Gamma(-s)(-z)^s \, ds}{\Gamma(\gamma-\alpha)\Gamma(\gamma-\beta)}$$

Change the order of integration, and use the result in Corollary 5.3. We get

$$2\pi i \frac{\Gamma(\alpha)\Gamma(\beta)\Gamma(\gamma-\alpha)\Gamma(\gamma-\beta)}{\Gamma(\gamma)} F(\alpha, \beta; \gamma; z)$$

$$= \int_B \left\{ \Gamma(\alpha+t)\Gamma(\beta+t)\Gamma(\gamma-\alpha-\beta-t) \underbrace{\left\{ \frac{1}{2\pi i} \int_L \Gamma(s-t)\Gamma(-s)(-z)^s \, ds \right\}}_{\Gamma(-t)(1-z)^t, \quad |z| < 1} dt \right\}$$

or

$$\frac{\Gamma(\alpha)\Gamma(\beta)\Gamma(\gamma-\alpha)\Gamma(\gamma-\beta)}{\Gamma(\gamma)} F(\alpha, \beta; \gamma; z)$$

$$= \frac{1}{2\pi i} \int_B \Gamma(\alpha+t)\Gamma(\beta+t)\Gamma(\gamma-\alpha-\beta-t)\Gamma(-t)(1-z)^t \, dt$$

We proceed by using the same arguments as in the proof of Barnes' lemma, Lemma 5.6. The integral on the right-hand side is evaluated in terms of the hypergeometric series by closing the contour in the right half plane and using the residue theorem and (5.15), i.e.,

$$\frac{\Gamma(\alpha)\Gamma(\beta)\Gamma(\gamma-\alpha)\Gamma(\gamma-\beta)}{\Gamma(\gamma)} F(\alpha, \beta; \gamma; z)$$

$$= \sum_{n=0}^{\infty} \Gamma(\alpha+n)\Gamma(\beta+n)\Gamma(\gamma-\alpha-\beta-n) \frac{(-1)^n}{n!}(1-z)^n$$

$$+ \sum_{n=0}^{\infty} \Gamma(\gamma-\alpha+n)\Gamma(\gamma-\beta+n)\Gamma(\alpha+\beta-\gamma-n) \frac{(-1)^n}{n!}(1-z)^{\gamma-\alpha-\beta+n}$$

By employing (A.10), we get

$$\frac{\Gamma(\alpha)\Gamma(\beta)\Gamma(\gamma-\alpha)\Gamma(\gamma-\beta)}{\Gamma(\gamma)} F(\alpha, \beta; \gamma; z)$$

$$= \sum_{n=0}^{\infty} \frac{\Gamma(\alpha+n)\Gamma(\beta+n)\Gamma(\gamma-\alpha-\beta)}{(1-\gamma+\alpha+\beta, n)} \frac{(1-z)^n}{n!}$$

$$+ (1-z)^{\gamma-\alpha-\beta} \sum_{n=0}^{\infty} \frac{\Gamma(\gamma-\alpha+n)\Gamma(\gamma-\beta+n)\Gamma(\alpha+\beta-\gamma)}{(1+\gamma-\alpha-\beta, n)} \frac{(1-z)^n}{n!}$$

which we rewrite as (5.12), viz.,

$$F(\alpha,\beta;\gamma;z) = \frac{\Gamma(\gamma)\Gamma(\gamma-\alpha-\beta)}{\Gamma(\gamma-\alpha)\Gamma(\gamma-\beta)}F(\alpha,\beta;1-\gamma+\alpha+\beta;1-z)$$
$$+ \frac{\Gamma(\gamma)\Gamma(\alpha+\beta-\gamma)}{\Gamma(\alpha)\Gamma(\beta)}(1-z)^{\gamma-\alpha-\beta}F(\gamma-\alpha,\gamma-\beta;1+\gamma-\alpha-\beta;1-z)$$

This result shows the nature of the singularity at $z = 1$, and also gives the linear relationship between solutions $u_1(z)$, $u_5(z)$, and $u_6(z)$ in Table 5.2.

5.7 Quadratic transformations

In Section 5.4, we investigated the consequences of the symmetry properties of the hypergeometric differential equation under the Möbius transformation. As a result, 24 different solutions were constructed from the hypergeometric function, see Table 5.2 on page 74. Additional symmetry properties are investigated in this section.

The roots to the indicial equation at $z = 0$ are 0 and $\gamma - 1$. One solution, the hypergeometric function, $F(\alpha,\beta;\gamma;z)$, is analytic at the origin, and the other solution has an exponent $\gamma - 1$ at the origin, which for non-integer values of γ leads to a branch point at the origin.[24] For special values of the indices α, β, and γ, the singularity at the origin can be removed. Specifically, let $\alpha = 2a$, $\beta = 2b$, and $\gamma = a+b+\frac{1}{2}$, i.e., the hypergeometric differential equation, see (4.3),

$$z(z-1)u''(z) + \left[(2a+2b+1)z - \left(a+b+\frac{1}{2}\right)\right]u'(z) + 4abu(z) = 0$$

Two linearly independent solutions of this equation are, see solutions $u_1(z)$ and $u_2(z)$ in Table 5.2 on page 74,

$$\begin{cases} F\left(2a, 2b; a+b+\frac{1}{2};z\right) \\ z^{\frac{1}{2}-a-b}F\left(a-b+\frac{1}{2}, b-a+\frac{1}{2}; \frac{3}{2}-a-b;z\right) \end{cases} \tag{5.16}$$

We notice that these solutions are identical if $\gamma = a+b+\frac{1}{2} = 1$, so this case must be excluded below. Make a change of the independent variable $t = 4z(1-z)$, with inverse $z = g(t) = (1-(1-t)^{1/2})/2$. The point $z = 0$ corresponds to $t = 0$. By the use of Theorem 2.1 on page 4, the new differential equation reads

$$t(t-1)v''(t) + \left[(a+b+1)t - \left(a+b+\frac{1}{2}\right)\right]v'(t) + abv(t) = 0$$

---

[24] We exclude the non-positive integers $\gamma = 0, -1, -2, \ldots$.

which again we recognize as the hypergeometric differential equation with indices $\alpha = a$, $\beta = b$, and $\gamma = a+b+\frac{1}{2}$. The solution that is analytic at the origin $t = 0$ is

$$F\left(a,b;a+b+\frac{1}{2};t\right) = F\left(a,b;a+b+\frac{1}{2};4z(1-z)\right)$$

This solution can be expressed as a linear combination of the two linear independent solutions in (5.16). However, the linear combination must be analytic at the origin $z = 0$, which implies that it is a multiple of the the first solution $F(2a,2b;a+b+1/2;z)$ alone. The multiplicative coefficient, however, must be 1, since both $F(a,b;a+b+1/2;4z(1-z))$ and $F(2a,2b;a+b+1/2;z)$ have the value 1 at the origin. We, therefore, conclude

$$F\left(a,b;a+b+\frac{1}{2};4z-4z^2\right) = F\left(2a,2b;a+b+\frac{1}{2};z\right)$$

or stated differently, with a shift in the argument $(z \rightarrow (1-z)/2$, $a \rightarrow a/2$ and $b \rightarrow b/2)$

$$F\left(\frac{a}{2},\frac{b}{2};\frac{a+b+1}{2};1-z^2\right) = F\left(a,b;\frac{a+b+1}{2};\frac{1-z}{2}\right)$$

Employing the Euler result, Lemma 5.3 on page 79, we have proved the following lemma:

Lemma 5.7. *The hypergeometric function satisfies*

$$F\left(a,b;\frac{a+b+1}{2};\frac{1-z}{2}\right) = z^{-a}F\left(\frac{a}{2},\frac{a+1}{2};\frac{a+b+1}{2};1-\frac{1}{z^2}\right)$$

This result is used below in the construction of a linearly independent solution of the Legendre[25] functions in Section 6.1.

5.8 Hypergeometric polynomials (Jacobi)

We have already noted that if α or β is a non-positive integer $-n$, $n \in \mathbb{N}$, the hypergeometric series $F(\alpha,\beta;\gamma;z)$ is terminated after a finite number of terms — a polynomial of degree n. In this section, we exploit the features of these polynomials in more detail.

---

[25] Adrien-Marie Legendre (1752–1833), French mathematician.

5.8.1 Definition of the Jacobi polynomials

To initiate the construction of the Jacobi[26] polynomials, we start by investigating the derivatives of

$$f_n(z) = z^{\alpha+n}(1-z)^{\beta+n}$$

for arbitrary real constants $\alpha > -1$ and $\beta > -1$ and n non-negative integer.[27] The reason for picking the points $z = 0, 1$ as special is that we want to identify the result below with regular singular points of the hypergeometric function.

The first derivative w.r.t. z is

$$f_n'(z) = \frac{\alpha+n}{z} f_n(z) - \frac{\beta+n}{1-z} f_n(z)$$

or

$$z(1-z)f_n'(z) = (\alpha+n)(1-z)f_n(z) - (\beta+n)zf_n(z)$$
$$= [\alpha+n-(2n+\alpha+\beta)z]f_n(z)$$

Repeated differentiation w.r.t. z using Leibniz' rule[28] for higher derivatives of a product[29] gives

$$z(1-z)f_n^{(m+1)}(z) + \binom{m}{1}(1-2z)f_n^{(m)}(z) + \binom{m}{2}(-2)f_n^{(m-1)}(z)$$

$$= [\alpha+n-(2n+\alpha+\beta)z]f_n^{(m)}(z) - \binom{m}{1}(2n+\alpha+\beta)f_n^{(m-1)}(z)$$

Let $m = n+1$ and introduce the notation $g_n(z) = f_n^{(n)}(z)$. We have

$$z(1-z)g_n''(z) + (n+1)(1-2z)g_n'(z) - n(n+1)g_n(z)$$
$$= [\alpha+n-(2n+\alpha+\beta)z]g_n'(z) - (n+1)(2n+\alpha+\beta)g_n(z)$$

or

$$z(z-1)g_n''(z) + [(2-\alpha-\beta)z+\alpha-1]g_n'(z) - (n+1)(n+\alpha+\beta)g_n(z) = 0$$

This is the hypergeometric differential equation

$$z(z-1)u''(z) + [(a+b+1)z-c]u'(z) + abu(z) = 0$$

---

[26] Carl Gustav Jacob Jacobi (1804–1851), Prussian mathematician.

[27] Generalizations to complex values of α and β satisfying $\mathrm{Re}\,\alpha > -1$ and $\mathrm{Re}\,\beta > -1$ exist.

[28] Gottfried Wilhelm Leibniz (1646–1716), German mathematician.

[29] Leibniz' rule for higher derivatives of a product reads

$$\frac{d^m(u(z)v(z))}{dz^m} = \sum_{k=0}^{m} \binom{m}{k} u^{(m-k)}(z)v^{(k)}(z)$$

with coefficients

$$\begin{cases} a+b=1-\alpha-\beta \\ ab=-(n+1)(n+\alpha+\beta) \\ c=1-\alpha \end{cases} \Rightarrow \begin{cases} a=n+1 \\ b=-n-\alpha-\beta \ \text{(or a and b interchanged)} \\ c=1-\alpha \end{cases}$$

Therefore

$$g_n(z) \in P \left\{ \begin{matrix} 0 & 1 & \infty \\ 0 & 0 & a \\ 1-c & c-a-b & b \end{matrix} \ z \right\} = P \left\{ \begin{matrix} 0 & 1 & \infty \\ 0 & 0 & n+1 \\ \alpha & \beta & -n-\alpha-\beta \end{matrix} \ z \right\}$$

Use the displacement theorem, see, e.g., (5.5) on page 68,

$$g_n(z) = \frac{d^n}{dz^n}\left(z^{\alpha+n}(1-z)^{\beta+n}\right) \in z^{\alpha}(1-z)^{\beta} P \left\{ \begin{matrix} 0 & 1 & \infty \\ 0 & 0 & n+\alpha+\beta+1 \\ -\alpha & -\beta & -n \end{matrix} \ z \right\}$$

This relation tells us that

$$z^{-\alpha}(1-z)^{-\beta} \frac{d^n}{dz^n}\left(z^{\alpha+n}(1-z)^{\beta+n}\right) \tag{5.17}$$

are polynomials of degree n in z, since one of the first coefficients in the hypergeometric function is a non-positive integer $-n$. This conclusion can, of course, also be made by a direct differentiation of the expression.

The polynomials constructed above lead us to the definition of the Jacobi polynomials, $P_n^{(\alpha,\beta)}(x)$ [11, 26]. The definition in terms of the hypergeometric function is

$$P_n^{(\alpha,\beta)}(x) = \binom{n+\alpha}{n} F\left(-n, n+\alpha+\beta+1; \alpha+1; \frac{1-x}{2}\right) \tag{5.18}$$

Notice that a shift in the argument has been made, so that instead of focusing on the point $z = 0, 1$, the focus is on $x = \pm 1$. Notice also that in the definition of the Jacobi polynomials we denote the independent variable x instead of the usual z. This is due to the fact that in most cases the argument is real-valued, e.g., $x \in [-1, 1]$. A series of special polynomials that occur in mathematical physics are in fact special cases of the Jacobi polynomials, see Table 5.4.

The definition also implies, see (A.14),

$$P_n^{(\alpha,\beta)}(1) = \binom{n+\alpha}{n} = \frac{\Gamma(n+\alpha+1)}{n!\Gamma(\alpha+1)} = \frac{(\alpha+1,n)}{n!} \tag{5.19}$$

The hypergeometric series in Theorem 5.1 provides an explicit form of the Jacobi polynomials, i.e.,

$$P_n^{(\alpha,\beta)}(x) = \binom{n+\alpha}{n} \sum_{k=0}^{n} \frac{(-n,k)(n+\alpha+\beta+1,k)}{(\alpha+1,k)k!} \left(\frac{1-x}{2}\right)^k$$

Table 5.4 Different polynomials that are special cases of the Jacobi polynomials when $\alpha = \beta$.

$\alpha = \beta$	Polynomial
	Gegenbauer, $G_n^\alpha(x)$
0	Legendre, $P_n(x)$
$-1/2$	Tchebysheff of the first kind, $T_n(\cos\theta) = \cos n\theta$
$1/2$	Tchebysheff of the second kind, $U_n(\cos\theta) = \sin((n+1)\theta)/\sin\theta$

or with the use of the binomial coefficient, see (A.14) on page 176, and

$$(-n,k) = (-n) \cdot \ldots \cdot (-n+k-1) = (-1)^k n \cdot \ldots \cdot (n+1-k) = (-1)^k \frac{n!}{(n-k)!}$$

we obtain

$$P_n^{(\alpha,\beta)}(x) = \frac{\Gamma(n+\alpha+1)}{n!} \sum_{k=0}^{n} \binom{n}{k} \frac{(n+\alpha+\beta+1,k)}{\Gamma(\alpha+k+1)} \left(\frac{x-1}{2}\right)^k \tag{5.20}$$

Moreover, the Jacobi polynomials satisfy

$$(1-x^2)u''(x) + [\beta - \alpha - (\alpha+\beta+2)x]u'(x) + n(n+\alpha+\beta+1)u(x) = 0 \tag{5.21}$$

where we used Theorem 2.1 on page 4, and $u(x) = P_n^{(\alpha,\beta)}(x)$.

By the use of Gauss' formula, see Corollary 5.1 on page 78, and (A.4),

$$F(-n, n+\alpha+\beta+1; \alpha+1; 1) = \frac{\Gamma(\alpha+1)\Gamma(-\beta)}{\Gamma(\alpha+1+n)\Gamma(-n-\beta)}$$

$$= (-1)^n \frac{\Gamma(\alpha+1)\Gamma(\beta+n+1)}{\Gamma(\alpha+1+n)\Gamma(\beta+1)}$$

we also have an explicit expression of the value of the Jacobi polynomials at $x = -1$, i.e.,

$$P_n^{(\alpha,\beta)}(-1) = (-1)^n \frac{\Gamma(\beta+n+1)}{n!\Gamma(\beta+1)} = (-1)^n \frac{(\beta+1,n)}{n!} \tag{5.22}$$

Using Lemma 5.3 on page 79, an alternative definition of the Jacobi polynomials is obtained

$$P_n^{(\alpha,\beta)}(x) = \left(\frac{1+x}{2}\right)^n \binom{n+\alpha}{n} F\left(-n, -n-\beta; \alpha+1; \frac{x-1}{x+1}\right)$$

The coefficient $c_n^{(\alpha,\beta)}$ of the highest power, x^n, in $P_n^{(\alpha,\beta)}(x)$ is used below. Since $P_n^{(\alpha,\beta)}(x)$ is a polynomial of degree n, this coefficient is determined by the use of (5.20). Only the highest power term in the hypergeometric series gives a contribution to the limit

$$c_n^{(\alpha,\beta)} = \frac{(n+\alpha+\beta+1,n)}{2^n n!} \tag{5.23}$$

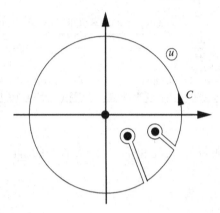

Fig. 5.7 The Contour C in the complex u-plane encircling the origin, but excluding the other two singularities of the integrand, $1/z$ and $1/(z-1)$.

since

$$\lim_{x\to\infty} x^{-n}\left(\frac{x-1}{2}\right)^n = 2^{-n}$$

5.8.2 Generating function

We start with another useful integral representation, which is derived in Problem 5.6. The result is

$$F(-n,n+\alpha+\beta+1;\alpha+1;z)$$
$$= \frac{n!\Gamma(\alpha+1)}{\Gamma(n+\alpha+1)}\frac{1}{2\pi i}\int_C \frac{du}{u^{n+1}}(1+(1-z)u)^{n+\alpha}(1-uz)^{n+\beta}$$

where the contour C encircles the singularity $u=0$, but not $u=1/z$ and $u=1/(z-1)$ in the complex u-plane, see Figure 5.7.

A simple change of variable $z=(1-x)/2$, $x\in[-1,1]$ gives an integral representation of the Jacobi polynomials that is a suitable start for finding the generating function of these polynomials. We get for $x\neq\pm1$ (we restrict the variable x to be real in this section)

$$P_n^{(\alpha,\beta)}(x) = \frac{1}{2\pi i}\int_C \frac{du}{u^{n+1}}\left(1+\frac{x+1}{2}u\right)^{n+\alpha}\left(1+\frac{x-1}{2}u\right)^{n+\beta}$$

Multiply with t^n and sum over n from $n=0$ to $n=\infty$. For sufficiently small values of t, we have

$$G^{(\alpha,\beta)}(x,t) \equiv \sum_{n=0}^{\infty} P_n^{(\alpha,\beta)}(x)t^n = \frac{1}{2\pi i}\int_C \frac{du}{u}\left(1+\frac{x+1}{2}u\right)^\alpha \left(1+\frac{x-1}{2}u\right)^\beta$$

$$\times \sum_{n=0}^{\infty}\left(1+\frac{x+1}{2}u\right)^n \left(1+\frac{x-1}{2}u\right)^n \left(\frac{t}{u}\right)^n$$

$$= \frac{1}{2\pi i}\int_C \left(1+\frac{x+1}{2}u\right)^\alpha \left(1+\frac{x-1}{2}u\right)^\beta \frac{du}{D}$$

where the denominator D is

$$D = u - \left(1+\frac{x+1}{2}u\right)\left(1+\frac{x-1}{2}u\right)t = -t\frac{x^2-1}{4}\left(u^2 - \frac{4u}{t}\frac{1-tx}{x^2-1}+\frac{4}{x^2-1}\right)$$

We seek the zeros of this denominator. The roots are

$$D=0 \quad\Longleftrightarrow\quad u = \frac{2}{t}\frac{1-tx}{x^2-1} \pm \frac{2}{t}\frac{R}{x^2-1} = \frac{2}{t}\frac{1-tx\pm R}{x^2-1} = u_\pm$$

where

$$R = \sqrt{1-2xt+t^2}$$

If we locate the position of the poles $u_\pm$, we see that

$$\begin{cases} u_+ = 2\dfrac{1-tx+\sqrt{1-2xt+t^2}}{t(x^2-1)} = 4\dfrac{1-tx+O(t^2)}{t(x^2-1)} = O(t^{-1}) \\[4mm] u_- = 2\dfrac{1-tx-\sqrt{1-2xt+t^2}}{t(x^2-1)} = 2\dfrac{O(t^2)}{t(x^2-1)} = O(t) \end{cases}$$

For sufficiently small t, only the pole $u = u_-$ lies inside the contour C in Figure 5.7, and only this pole contributes with its residue to the integral. The residue of the integrand at the poles $u = u_-$ is

$$\operatorname*{Res}_{u=u_-}\left(1+\frac{x+1}{2}u\right)^\alpha \left(1+\frac{x-1}{2}u\right)^\beta \frac{1}{D}$$

$$= -\left(1+\frac{1-tx-R}{t(x-1)}\right)^\alpha \left(1+\frac{1-tx-R}{t(x+1)}\right)^\beta \frac{4}{t(x^2-1)(u_- - u_+)}$$

We simplify

$$\begin{cases} 1+\dfrac{1-tx-R}{t(x-1)} = \dfrac{1-t-R}{t(x-1)} = \dfrac{2}{1-t+R} \\[4mm] 1+\dfrac{1-tx-R}{t(x+1)} = \dfrac{1+t-R}{t(x+1)} = \dfrac{2}{1+t+R} \end{cases}$$

and, finally, we get

$$G^{(\alpha,\beta)}(x,t) = \left(\frac{2}{1-t+R}\right)^\alpha \left(\frac{2}{1+t+R}\right)^\beta \frac{1}{R}$$

or

$$G^{(\alpha,\beta)}(x,t) = \sum_{n=0}^{\infty} P_n^{(\alpha,\beta)}(x)t^n = \frac{2^{\alpha+\beta}}{(1-t+R)^{\alpha}(1+t+R)^{\beta}R} \qquad (5.24)$$

Example 5.4. If $\alpha = \beta = 0$, we obtain the generating function for the Legendre[30] polynomials, $P_n(x)$, see Table 5.4.

$$G^{(0,0)}(x,t) = \sum_{n=0}^{\infty} P_n(x)t^n = \frac{1}{R} = \frac{1}{\sqrt{1-2xt+t^2}}$$

∎

Example 5.5. Evaluate the relation (5.24) for $-x$. This leads to

$$\sum_{n=0}^{\infty} P_n^{(\alpha,\beta)}(-x)t^n = \frac{2^{\alpha+\beta}}{(1-t+\widetilde{R})^{\alpha}(1+t+\widetilde{R})^{\beta}\widetilde{R}}$$

where

$$\widetilde{R} = \sqrt{1+2xt+t^2}$$

However, evaluating the relation (5.24) at $-t$ with α and β interchanged gives

$$\sum_{n=0}^{\infty} P_n^{(\beta,\alpha)}(x)(-t)^n = \frac{2^{\alpha+\beta}}{(1+t+\widetilde{R})^{\beta}(1-t+\widetilde{R})^{\alpha}\widetilde{R}}$$

which is identical to the above and we conclude

$$P_n^{(\alpha,\beta)}(-x) = (-1)^n P_n^{(\beta,\alpha)}(x)$$

∎

5.8.3 Rodrigues' generalized function

With the notation and the results of Section 5.8.1, we have (see (5.17), $z \to (1 - x)/2$)

$$P_n^{(\alpha,\beta)}(x) = C(1-x)^{-\alpha}(1+x)^{-\beta}\frac{d^n}{dx^n}\left((1-x)^{\alpha+n}(1+x)^{\beta+n}\right)$$

where the constant C can be determined by evaluating both sides at $x = 1$ using Leibniz' rule for differentiation of a product, see footnote 29 on page 93. The result is Rodrigues' generalized function,[31] see Problem 5.5.

$$P_n^{(\alpha,\beta)}(x) = \frac{(-1)^n}{2^n n!}(1-x)^{-\alpha}(1+x)^{-\beta}\frac{d^n}{dx^n}\left((1-x)^{\alpha+n}(1+x)^{\beta+n}\right) \qquad (5.25)$$

---

[30] Adrien-Marie Legendre (1752–1833), French mathematician.
[31] Benjamin Olinde Rodrigues (1795–1851), French mathematician.

Example 5.6. If $\alpha = \beta = 0$, we obtain Rodrigues' generalized function for the Legendre polynomials, $P_n(x)$, see Table 5.4.

$$P_n(x) = \frac{1}{2^n n!} \frac{d^n}{dx^n} \left(x^2 - 1\right)^n$$

∎

5.8.4 Orthogonality

The Jacobi polynomials, $P_n^{(\alpha,\beta)}(x)$, are orthogonal over the interval $[-1,1]$ with the weight function $(1-x)^\alpha (1+x)^\beta$ (α and β are assumed real constants larger than -1, and n a non-negative integer). To see this, define

$$I_{n,m} = \int_{-1}^{1} (1-x)^\alpha (1+x)^\beta P_n^{(\alpha,\beta)}(x) P_m^{(\alpha,\beta)}(x)\, dx$$

$$= \frac{(-1)^n}{2^n n!} \int_{-1}^{1} \frac{d^n}{dx^n} \left((1-x)^{\alpha+n}(1+x)^{\beta+n}\right) P_m^{(\alpha,\beta)}(x)\, dx$$

where (5.25) is introduced. If $m < n$, then from integration by parts n times, we obtain (there is no loss of generality to assume that $m < n$, since $I_{n,m}$ is symmetric in n and m)

$$I_{n,m} = \frac{1}{2^n n!} \int_{-1}^{1} (1-x)^{\alpha+n}(1+x)^{\beta+n} \frac{d^n}{dx^n} P_m^{(\alpha,\beta)}(x)\, dx = 0$$

since $P_m^{(\alpha,\beta)}(x)$ is a polynomial of degree $m < n$. If $m = n$, we again have from integration by parts n times

$$I_{n,n} = \frac{1}{2^n n!} \int_{-1}^{1} (1-x)^{\alpha+n}(1+x)^{\beta+n} \frac{d^n}{dx^n} P_n^{(\alpha,\beta)}(x)\, dx$$

Since $P_n^{(\alpha,\beta)}(x)$ is a polynomial of degree n, with the coefficient $c_n^{(\alpha,\beta)}$ in front of the highest power x^n, see (5.23), we get

$$I_{n,n} = \frac{c_n^{(\alpha,\beta)}}{2^n} \int_{-1}^{1} (1-x)^{\alpha+n}(1+x)^{\beta+n}\, dx = \frac{c_n^{(\alpha,\beta)}}{2^{n-1}} \int_{0}^{1} (2-2t)^{\alpha+n}(2t)^{\beta+n}\, dt$$

$$= 2^{\alpha+\beta+n+1} c_n^{(\alpha,\beta)} \frac{\Gamma(\alpha+n+1)\Gamma(\beta+n+1)}{\Gamma(\alpha+\beta+2n+2)}$$

by the beta function (A.16) and (A.17). With (5.23) we finally get

$$I_{n,n} = \frac{2^{\alpha+\beta+1}}{\alpha+\beta+2n+1} \frac{\Gamma(\alpha+n+1)\Gamma(\beta+n+1)}{n!\,\Gamma(\alpha+\beta+n+1)}$$

The orthogonality is summarized in the following lemma:

Lemma 5.8. *The Jacobi polynomials,* $P_n^{(\alpha,\beta)}(x)$, *are orthogonal over the interval* $[-1,1]$ *with the weight function* $(1-x)^\alpha(1+x)^\beta$, *where* α *and* β *are assumed real constants larger than* -1, *and* $n \in \mathbb{N}$. *The exact expression reads*

$$\int_{-1}^{1} (1-x)^\alpha (1+x)^\beta P_n^{(\alpha,\beta)}(x) P_m^{(\alpha,\beta)}(x)\, dx$$

$$= \frac{\delta_{n,m} 2^{\alpha+\beta+1}\Gamma(\alpha+n+1)\Gamma(\beta+n+1)}{n!(\alpha+\beta+2n+1)\Gamma(\alpha+\beta+n+1)}$$

Example 5.7. In particular, for the Legendre polynomials, $P_n(x)$, where $\alpha = \beta = 0$, we have the well-known normalization integral, i.e.,

$$\int_{-1}^{1} P_n(x) P_m(x)\, dx = \frac{2\delta_{n,m}}{2n+1}$$

∎

5.8.5 Integral representation (Schläfli)

Integral representations of the Jacobi polynomials are useful, and in this section we derive Schläfli's integral representation[32] of $P_n^{(\alpha,\beta)}(x)$.

Rodrigues' generalized function, (5.25), contains an nth order derivative, viz.,

$$\frac{d^n}{dx^n}\left((1-x)^{\alpha+n}(1+x)^{\beta+n} \right)$$

By the use of, see footnote 10 on page 19,

$$f^{(n)}(x) = \frac{n!}{2\pi i} \oint_C \frac{f(t)\, dt}{(t-x)^{n+1}}, \qquad n \in \mathbb{Z}_+$$

where the contour C encircles the point $t = x$ in a positive way, and where $f(t)$ is assumed analytic on and inside C, we obtain

$$\frac{d^n}{dx^n}\left((1-x)^{\alpha+n}(1+x)^{\beta+n} \right) = \frac{n!}{2\pi i} \oint_C \frac{(1-t)^{\alpha+n}(1+t)^{\beta+n}}{(t-x)^{n+1}}\, dt$$

where the contour C encircles the point $t = x$ in a positive way, but it does not encircle the points $t = \pm 1$, which are branch points of the integrand.

With Rodrigues' generalized function in (5.25), we have then proved the integral representation of the Jacobi polynomials. We summarize

---

[32] Ludwig Schläfli (1814–1895), Swiss mathematician.

Lemma 5.9. *The Schläfli's integral representation of* $P_n^{(\alpha,\beta)}(x)$ *is*

$$P_n^{(\alpha,\beta)}(x) = \frac{2^{-n}}{2\pi i} \oint_C (t^2-1)^n \left(\frac{1-t}{1-x}\right)^\alpha \left(\frac{1+t}{1+x}\right)^\beta (t-x)^{-n-1} \, dt$$

where the contour C encircles the point $t = x$ *in a positive way, but it does not encircle the points* $t = \pm 1$.

Example 5.8. In particular, for the Legendre polynomials, $P_n(x)$, where $\alpha = \beta = 0$, we get

$$P_n(x) = \frac{2^{-n}}{2\pi i} \oint_C \frac{(t^2-1)^n}{(t-x)^{n+1}} \, dt \qquad (5.26)$$

■

5.8.6 Recursion relation

Between any three consecutive orthogonal Jacobi polynomials there is a relation

$$P_{n+1}^{(\alpha,\beta)}(x) = (A_n^{(\alpha,\beta)}x + B_n^{(\alpha,\beta)})P_n^{(\alpha,\beta)}(x) + C_n^{(\alpha,\beta)}P_{n-1}^{(\alpha,\beta)}(x) \qquad (5.27)$$

for some constants $A_n^{(\alpha,\beta)}$, $B_n^{(\alpha,\beta)}$, and $C_n^{(\alpha,\beta)}$. To see this, determine $A_n^{(\alpha,\beta)}$, the exact value is given below, such that $P_{n+1}^{(\alpha,\beta)}(x) - A_n^{(\alpha,\beta)}xP_n^{(\alpha,\beta)}(x)$ is a polynomial of degree n. This is always possible by adjusting the value of $A_n^{(\alpha,\beta)}$ so that the terms of order $n+1$ cancel. We write the right-hand side as a sum of Jacobi polynomials of degree less than or equal to n, i.e.,

$$P_{n+1}^{(\alpha,\beta)}(x) - A_n^{(\alpha,\beta)}xP_n^{(\alpha,\beta)}(x) = \sum_{k=0}^n a_k^{(\alpha,\beta)}P_k^{(\alpha,\beta)}(x) \qquad (5.28)$$

However, there are at most two non-vanishing terms ($k = n, n-1$) in the series on the right-hand side, since by orthogonality, see Lemma 5.8,

$$\int_{-1}^1 P_n^{(\alpha,\beta)}(x)P_m^{(\alpha,\beta)}(x)w^{(\alpha,\beta)}(x) \, dx = N_n^{(\alpha,\beta)}\delta_{n,m}$$

where

$$N_n^{(\alpha,\beta)} = \frac{2^{\alpha+\beta+1}\Gamma(\alpha+n+1)\Gamma(\beta+n+1)}{n!(\alpha+\beta+2n+1)\Gamma(\alpha+\beta+n+1)}, \qquad w^{(\alpha,\beta)}(x) = (1-x)^\alpha(1+x)^\beta$$

The orthonormal properties and (5.28) give

$$N_k^{(\alpha,\beta)} a_k^{(\alpha,\beta)} = -A_n^{(\alpha,\beta)} \int_{-1}^{1} P_n^{(\alpha,\beta)}(x) \underbrace{x P_k^{(\alpha,\beta)}(x)}_{\sum_{l=0}^{k+1} \lambda_l^{(\alpha,\beta)} P_l^{(\alpha,\beta)}(x)} w^{(\alpha,\beta)}(x)\, dx = 0, \quad \text{if } k < n-1$$

This identity implies that $a_k^{(\alpha,\beta)} = 0$, if $k < n-1$. The existence of a recursion relation of the kind given in (5.27) is therefore proved.

We proceed by determining the coefficients $A_n^{(\alpha,\beta)}$, $B_n^{(\alpha,\beta)}$, and $C_n^{(\alpha,\beta)}$ in (5.27). The coefficients in front of the nth power of x in the Jacobi polynomial $P_n^{(\alpha,\beta)}(x)$ is, see (5.23),

$$c_n^{(\alpha,\beta)} = \frac{(n+\alpha+\beta+1,n)}{2^n n!}$$

Therefore, since the highest powers in x on both sides in the recursion relation (5.27) match, we have using (A.9)

$$A_n^{(\alpha,\beta)} = \frac{c_{n+1}^{(\alpha,\beta)}}{c_n^{(\alpha,\beta)}} = \frac{(n+\alpha+\beta+2,n+1)}{2(n+1)(n+\alpha+\beta+1,n)} = \frac{(2n+\alpha+\beta+1)(2n+\alpha+\beta+2)}{2(n+1)(n+\alpha+\beta+1)}$$

The other two coefficients, $B_n^{(\alpha,\beta)}$ and $C_n^{(\alpha,\beta)}$, are determined by the use of the orthogonality relation and the explicit value at $x = 1$. We get by orthogonality

$$N_{n-1}^{(\alpha,\beta)} C_n^{(\alpha,\beta)} = -A_n^{(\alpha,\beta)} \int_{-1}^{1} P_n^{(\alpha,\beta)}(x)\, x \underbrace{P_{n-1}^{(\alpha,\beta)}(x)}_{c_{n-1}^{(\alpha,\beta)} x^n + \ldots}\, w^{(\alpha,\beta)}(x)\, dx$$

$$= -A_n^{(\alpha,\beta)} \frac{c_{n-1}^{(\alpha,\beta)}}{c_n^{(\alpha,\beta)}} N_n^{(\alpha,\beta)} = -\frac{A_n^{(\alpha,\beta)}}{A_{n-1}^{(\alpha,\beta)}} N_n^{(\alpha,\beta)}$$

from which we can solve for $C_n^{(\alpha,\beta)}$,

$$C_n^{(\alpha,\beta)} = -\frac{A_n^{(\alpha,\beta)} N_n^{(\alpha,\beta)}}{A_{n-1}^{(\alpha,\beta)} N_{n-1}^{(\alpha,\beta)}} = -\frac{(2n+\alpha+\beta+2)(\alpha+n)(\beta+n)}{(n+1)(n+\alpha+\beta+1)(2n+\alpha+\beta)}$$

By (5.19), we get

$$P_n^{(\alpha,\beta)}(1) = \frac{\Gamma(n+\alpha+1)}{n!\,\Gamma(\alpha+1)} = \frac{(\alpha+1,n)}{n!}$$

The recursion relation (5.27), evaluated at $x = 1$, is (use (A.9))

$$\frac{(n+\alpha)(n+\alpha+1)}{n(n+1)} = (A_n^{(\alpha,\beta)} + B_n^{(\alpha,\beta)})\frac{(n+\alpha)}{n} + C_n^{(\alpha,\beta)}$$

which we solve for $B_n^{(\alpha,\beta)}$. The result is

$$B_n^{(\alpha,\beta)} = \frac{n+\alpha+1}{n+1} - A_n^{(\alpha,\beta)} - \frac{n}{(n+\alpha)}C_n^{(\alpha,\beta)}$$

All three coefficients, $A_n^{(\alpha,\beta)}$, $B_n^{(\alpha,\beta)}$, and $C_n^{(\alpha,\beta)}$, are then determined. A more compact form of the relation is obtained by identifying the common denominator of the coefficients $A_n^{(\alpha,\beta)}$, $B_n^{(\alpha,\beta)}$, and $C_n^{(\alpha,\beta)}$, and writing the recursion relation in the form

$$2(n+1)(n+\alpha+\beta+1)(2n+\alpha+\beta)P_{n+1}^{(\alpha,\beta)}(x)$$
$$= (a_n^{(\alpha,\beta)}x+b_n^{(\alpha,\beta)})P_n^{(\alpha,\beta)}(x) + c_n^{(\alpha,\beta)}P_{n-1}^{(\alpha,\beta)}(x)$$

where

$$\begin{cases} a_n^{(\alpha,\beta)} = 2(n+1)(n+\alpha+\beta+1)(2n+\alpha+\beta)A_n^{(\alpha,\beta)} \\ \qquad = (2n+\alpha+\beta,3) \\ c_n^{(\alpha,\beta)} = 2(n+1)(n+\alpha+\beta+1)(2n+\alpha+\beta)C_n^{(\alpha,\beta)} \\ \qquad = -2(\alpha+n)(\beta+n)(2n+\alpha+\beta+2) \end{cases}$$

and

$$b_n^{(\alpha,\beta)} = 2(n+1)(n+\alpha+\beta+1)(2n+\alpha+\beta)\frac{n+\alpha+1}{n+1} - a_n^{(\alpha,\beta)} - \frac{n}{(n+\alpha)}c_n^{(\alpha,\beta)}$$
$$= 2(n+\alpha+\beta+1)(2n+\alpha+\beta)(n+\alpha+1) - (2n+\alpha+\beta,3)$$
$$+ 2n(\beta+n)(2n+\alpha+\beta+2) = (2n+\alpha+\beta+1)(\alpha^2-\beta^2)$$

We have thus proved the next lemma.

Lemma 5.10. *The recursion relation between three consecutive Jacobi polynomials reads* $(n \in \mathbb{Z}_+)$

$$2(n+1)(n+\alpha+\beta+1)(2n+\alpha+\beta)P_{n+1}^{(\alpha,\beta)}(x)$$
$$= ((2n+\alpha+\beta+1)(\alpha^2-\beta^2)+(2n+\alpha+\beta,3)x)\,P_n^{(\alpha,\beta)}(x)$$
$$- 2(n+\alpha)(n+\beta)(2n+\alpha+\beta+2)P_{n-1}^{(\alpha,\beta)}(x)$$

The first two Jacobi polynomials, $P_0^{(\alpha,\beta)}(x)$ and $P_1^{(\alpha,\beta)}(x)$, are easily found by explicitly evaluating the hypergeometric series or using (5.25). They are

$$\begin{cases} P_0^{(\alpha,\beta)}(x) = 1 \\ P_1^{(\alpha,\beta)}(x) = \frac{1}{2}(\alpha+\beta+2)x+\frac{1}{2}(\alpha-\beta) \end{cases}$$

These two Jacobi polynomials are used to initialize the recursion relation. Indeed, the recursion relation in Lemma 5.10 holds for $n=0$ provided we make the additional definition $P_{-1}^{(\alpha,\beta)}(x) = 0$, since

$$2(\alpha+\beta+1)(\alpha+\beta)\underbrace{\left(\frac{1}{2}(\alpha+\beta+2)x+\frac{1}{2}(\alpha-\beta)\right)}_{P_1^{(\alpha,\beta)}(x)}$$

$$=(\alpha+\beta+1)(\alpha^2-\beta^2)+(\alpha+\beta,3)x$$

The recursion relation can be used to obtain integrals of products of Jacobi polynomials in combination with x. If the recursion relation in Lemma 5.10 is multiplied by $P_k^{(\alpha,\beta)}(x)w^{(\alpha,\beta)}(x)$ and integrated over $x \in [-1,1]$, the orthogonality relation, see Lemma 5.8, implies

$$2(n+1)(n+\alpha+\beta+1)(2n+\alpha+\beta)N_{n+1}^{(\alpha,\beta)}\delta_{k,n+1}$$

$$=(2n+\alpha+\beta+1)(\alpha^2-\beta^2)N_n^{(\alpha,\beta)}\delta_{k,n}$$

$$+(2n+\alpha+\beta,3)\int_{-1}^1 xP_n^{(\alpha,\beta)}(x)P_k^{(\alpha,\beta)}(x)w^{(\alpha,\beta)}(x)\,dx$$

$$-2(n+\alpha)(n+\beta)(2n+\alpha+\beta+2)N_{n-1}^{(\alpha,\beta)}\delta_{k,n-1}$$

from which we can find the solution of the integral

$$I_{k,n}=\int_{-1}^1 xP_n^{(\alpha,\beta)}(x)P_k^{(\alpha,\beta)}(x)w^{(\alpha,\beta)}(x)\,dx$$

$$=\frac{2(n+1)(n+\alpha+\beta+1)}{(2n+\alpha+\beta+1)(2n+\alpha+\beta+2)}N_{n+1}^{(\alpha,\beta)}\delta_{k,n+1}$$

$$-\frac{(\alpha^2-\beta^2)}{(2n+\alpha+\beta)(2n+\alpha+\beta+2)}N_n^{(\alpha,\beta)}\delta_{k,n}$$ (5.29)

$$+\frac{2(n+\alpha)(n+\beta)}{(2n+\alpha+\beta)(2n+\alpha+\beta+1)}N_{n-1}^{(\alpha,\beta)}\delta_{k,n-1}$$

Example 5.9. In particular, for the Legendre polynomials, $P_n(x)$, where $\alpha=\beta=0$, the integrals in (5.29) become

$$I_{k,n}=\int_{-1}^1 xP_n(x)P_k(x)\,dx=\frac{2(n+1)}{(2n+1)(2n+3)}\delta_{k,n+1}+\frac{2n}{(2n-1)(2n+1)}\delta_{k,n-1}$$

∎

Problems

5.1. Prove the sum for $a,b,c \in \mathbb{C}$, $c \notin \mathbb{Z}_-$

$$\sum_{k=0}^{n} \frac{(a+c,k)(b+c,k)}{(c+1,k)k!} \{(a+b+c)(c+k)+ab\}$$
$$= \frac{(a+c,n+1)(b+c,n+1)}{(c+1,n)n!}, \quad n \in \mathbb{N}$$

Hint: Combine the results in (5.1) with (5.4) or use Table 5.2, and use the coefficients a_n' in (2.15).

5.2. Prove

$$\frac{d}{dz} F(\alpha,\beta;\gamma;z) = \frac{\alpha\beta}{\gamma} F(\alpha+1,\beta+1;\gamma+1;z)$$

and more generally

$$\frac{d^n}{dz^n} F(\alpha,\beta;\gamma;z) = \frac{(\alpha,n)(\beta,n)}{(\gamma,n)} F(\alpha+n,\beta+n;\gamma+n;z)$$

5.3. For $\alpha - \beta$ not a negative integer and α not an even negative integer, show

$$F(\alpha,\beta;1+\alpha-\beta;-1) = \frac{\Gamma(1+\alpha-\beta)\Gamma(1+\alpha/2)}{\Gamma(1+\alpha/2-\beta)\Gamma(1+\alpha)}$$

5.4. Show that

$$F(\alpha,\beta;\beta;z) = (1-z)^{-\alpha}, \quad z \neq 1$$

5.5. Show that the normalization constant C in Rodrigues' generalized function, see Section 5.8.3, is

$$C = \frac{(-1)^n}{2^n n!}$$

5.6. [†]Prove, for non-negative integer values n,

$$F(-n,n+\alpha+\beta+1;\alpha+1;z)$$
$$= \frac{n!\Gamma(\alpha+1)}{\Gamma(n+\alpha+1)} \frac{1}{2\pi i} \int_C \frac{du}{u^{n+1}} (1+(1-z)u)^{n+\alpha} (1-uz)^{n+\beta}$$

where the contour C encircles the singularity $u=0$, but not $u=1/z$ and $u=1/(z-1)$ in the complex u-plane, see Figure 5.7.
Hint: Use the Euler relation in Lemma 5.3.

5.7. [†]Verify the recursion relation for the Jacobi polynomials that also contains a derivative of the Jacobi polynomials ($n \in \mathbb{Z}_+$)

$$(2n+\alpha+\beta)(1-x^2)\frac{d}{dx} P_n^{(\alpha,\beta)}(x)$$
$$= n(\alpha-\beta-(2n+\alpha+\beta)x) P_n^{(\alpha,\beta)}(x) + 2(n+\alpha)(n+\beta)P_{n-1}^{(\alpha,\beta)}(x)$$

Does the relation hold for $n=0$ and $n=1$?

Chapter 6
Legendre functions and related functions

Due to the great importance of the Legendre functions in mathematical physics, we devote a whole chapter to analyze these functions, and the functions that are related to the Legendre functions, in detail. As an important tool, we use the integral representations developed in Chapter 5.

6.1 Legendre functions of first and second kind

In Section 5.8, we investigated the properties of polynomial solutions of the hypergeometric equation. Specifically, we found that the Legendre polynomials,[1] $P_n(z)$, satisfy, see Table 5.4 on page 95 (use $\alpha = \beta = 0$) and (5.21) on page 95,

$$(1 - z^2)u''(z) - 2zu'(z) + n(n+1)u(z) = 0$$

The solution that is regular at the point $z = 1$ can be expressed in the hypergeometric function $F(\alpha, \beta; \gamma; z)$, see (5.18),

$$P_n(z) = F\left(-n, n+1; 1; \frac{1-z}{2}\right)$$

We now extend the solutions of this differential equation to non-integer values of n, and we use the index ν, which in general can be complex-valued, i.e., the Legendre differential equation reads

$$(z^2 - 1)P_\nu''(z) + 2zP_\nu'(z) - \nu(\nu+1)P_\nu(z) = 0 \tag{6.1}$$

and the Legendre function of the first kind is defined as

---

[1] The independent variable takes complex values in this chapter, in contrast to the case in Section 5.8, so we prefer to denote the independent variable z in this chapter.

G. Kristensson, *Second Order Differential Equations: Special Functions and Their Classification*, DOI 10.1007/978-1-4419-7020-6_6,
© Springer Science+Business Media, LLC 2010

$$P_v(z) = F\left(-v, v+1; 1; \frac{1-z}{2}\right) \qquad (6.2)$$

Note that

$$P_v(1) = 1, \quad \text{for all } v \in \mathbb{C}$$

and, moreover, $P_v(z)$ is an entire function in v for fixed z. The point $z = -1$ and the point at infinity are in general branch points, and to make the Legendre function single-valued, a branch cut along the negative real axis, from $z = -1$ to infinity, is introduced. We also conclude that

$$P_{-v-1}(z) = P_v(z), \quad \text{for all } v \in \mathbb{C}$$

since the order of the first two indices in the hypergeometric function is non-essential.

We continue by finding a second independent solution to the Legendre differential equation (6.1). This can be obtained by taking an independent solution in Table 5.2 on page 74. However, we follow the traditional way of finding a second solution to the Legendre differential equation due to Hobson.[2] This particular second solution, which is denoted $Q_v(z)$, is well-behaved at infinity.

We proceed by using Lemma 5.7 to find another, independent, way of expressing the Legendre function of the first kind. We get

$$P_v(z) = z^v F\left(-\frac{v}{2}, \frac{1-v}{2}; 1; 1 - \frac{1}{z^2}\right) \qquad (6.3)$$

With this result as a starting point, we construct another, linearly independent, solution to $P_v(z)$, by using the result from Section 5.4 that $u_1(z)$ and $u_6(z)$ are linearly independent solutions to the same equation. From Table 5.2 on page 74, we see that the functions

$$F\left(-\frac{v}{2}, \frac{1-v}{2}; 1; 1 - \frac{1}{z^2}\right) \text{ and } z^{-2v-1} F\left(1 + \frac{v}{2}, \frac{1+v}{2}; v + \frac{3}{2}; \frac{1}{z^2}\right)$$

satisfy the same differential equation. Another solution to the Legendre differential equation therefore is

$$z^{-v-1} F\left(1 + \frac{v}{2}, \frac{1+v}{2}; v + \frac{3}{2}; \frac{1}{z^2}\right)$$

which is linearly independent to the solution (6.3). From this result, we define the Legendre function of the second kind, $Q_v(z)$, as

$$Q_v(z) = \frac{\sqrt{\pi}}{(2z)^{v+1}} \frac{\Gamma(v+1)}{\Gamma(v+3/2)} F\left(1 + \frac{v}{2}, \frac{1+v}{2}; v + \frac{3}{2}; \frac{1}{z^2}\right) \qquad (6.4)$$

---

[2] Ernest William Hobson (1856–1933), English mathematician.

The regular singular points of the Legendre function of the second kind are the points at $z = \pm 1$ and at the origin.

The relation between the first and second kind solutions becomes more apparent if we use (6.3) and (5.12). The result is

$$
P_v(z) = \frac{\Gamma(v+\frac{1}{2})}{\Gamma(1+\frac{v}{2})\Gamma(\frac{1+v}{2})} z^v F(-\frac{v}{2}, \frac{1-v}{2}; \frac{1}{2} - v; \frac{1}{z^2})
$$
$$
+ \frac{\Gamma(-v-\frac{1}{2})}{\Gamma(\frac{-v}{2})\Gamma(\frac{1-v}{2})} z^{-v-1} F(1+\frac{v}{2}, \frac{1+v}{2}; v+\frac{3}{2}; \frac{1}{z^2})
$$

This equality can be simplified by the use of the properties of the gamma function, see (A.4) and (A.5). We get for $v \notin \mathbb{N}$

$$
P_v(z) = \frac{\Gamma(v+\frac{1}{2})}{\sqrt{\pi}\Gamma(v+1)} (2z)^v F(-\frac{v}{2}, \frac{1-v}{2}; \frac{1}{2} - v; \frac{1}{z^2})
$$
$$
+ \frac{\Gamma(-v-\frac{1}{2})}{\sqrt{\pi}\Gamma(-v)} (2z)^{-v-1} F(1+\frac{v}{2}, \frac{1+v}{2}; v+\frac{3}{2}; \frac{1}{z^2})
$$
$$
= \frac{\tan v\pi}{\pi} (Q_v(z) - Q_{-v-1}(z))
$$

where we also used (6.4). In summary,

$$
\pi \cot v\pi P_v(z) = Q_v(z) - Q_{-v-1}(z)
$$

6.2 Integral representations

Integral representations of Legendre functions of the first and second kind are proved useful in applications and can be used as alternative definitions of these functions. These integral representations are special cases of the more general integral representations of the hypergeometric function developed in Chapter 5.

An integral representation for the Legendre polynomials is given in (5.26)

$$
P_n(x) = \frac{2^{-n}}{2\pi i} \oint_C \frac{(t^2-1)^n}{(t-x)^{n+1}} dt
$$

where the contour C encircles the point $t = x$ in a positive way in the complex t-plane. This representation generalizes to non-integer values v of n, provided the contour also includes the point $t = 1$ but not the point $t = -1$, see Figure 6.1. The enclosure of the point $t = 1$ gives no extra contribution to the integral for integer order, $v = n$, but ensures that the integrand resumes its original value after encircling the contour C. In fact, the integrand

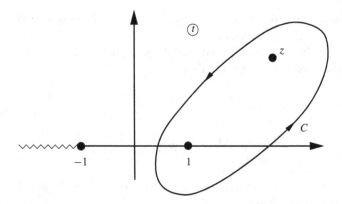

Fig. 6.1 The contour C in the complex t-plane used in Theorem 6.1. The complex t-plane has a branch cut along the negative real axis from -1 to $-\infty$.

$$\frac{\left(t^2 - 1\right)^{\nu}}{(t - z)^{\nu+1}}$$

resumes its original value multiplied by a factor $\exp\{-2i\pi(\nu + 1)\}$ after encircling the point $t = z$ in a counter-clockwise direction, and a factor $\exp\{2i\nu\pi\}$ after encircling the point $t = 1$. We summarize the result in a theorem.

Theorem 6.1 (Schläfli). *The Legendre function of the first kind, $P_{\nu}(z)$, has an integral representation*

$$P_{\nu}(z) = \frac{2^{-\nu}}{2\pi i} \oint_C \frac{\left(t^2 - 1\right)^{\nu}}{(t - z)^{\nu+1}}\, dt$$

where the contour C encircles the points $t = 1$ and $t = z$ counter-clockwise excluding the point $t = -1$, see Figure 6.1. It is assumed that the argument z is not on the branch cut along the negative real axis from -1 to $-\infty$, i.e., $z \notin (-\infty, -1]$.

Proof. We prove the theorem by showing that the contour integral in the theorem is identical to the representation of the hypergeometric function in Theorem 5.3 on page 77 for all $-1 < \mathrm{Re}\,\nu < 0$ and all $\mathbb{C} \ni z \notin (-\infty, -1]$. The representation in the theorem is then valid for all parameter values by analytic continuation, since both sides in the expression are well defined for all values of ν.

The integrand has three singular points $t = \pm 1$ and $t = z$. For general non-integer values of ν these points are branch points. We choose branch cuts along the line connecting $t = 1$ and $t = z$, and from $t = -1$ along the negative real t-axis, see Figure 6.2. We first conclude that the two circles around $t = 1$ and $t = z$ give vanishing contributions as the radii become small provided $-1 < \mathrm{Re}\,\nu < 0$.

On the right-hand side of the branch cut between the branch points $t = 1$ and $t = z$ we have

$$\begin{cases} t - 1 = |t - 1|e^{i\phi} \\ t - z = |t - z|e^{i\theta} \end{cases}$$

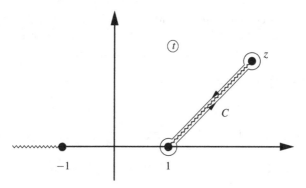

Fig. 6.2 The contour C in the complex t-plane along the branch cut between $t = 1$ and $t = z$.

where the angles are related to each other on the right-hand side of the branch cut as $\phi - \theta = \pi$, see Figure 6.3. On the left-hand side of the branch cut, encircling the point $t = z$, we have

$$\begin{cases} t - 1 = |t - 1| e^{i\phi} \\ t - z = |t - z| e^{i\theta + 2\pi i} \end{cases}$$

Notice also that the value of the integrand resumes its original value after encircling both branch points $t = 1$ and $t = z$, since both the numerator and the denominator differ by a factor $e^{2i\nu\pi}$.

The contour integral becomes

$$\oint_C \frac{(t^2 - 1)^\nu}{(t - z)^{\nu+1}} \, dt = \left(1 - e^{-2i\pi(\nu+1)}\right) \int_L \frac{(t^2 - 1)^\nu}{(t - z)^{\nu+1}} \, dt$$

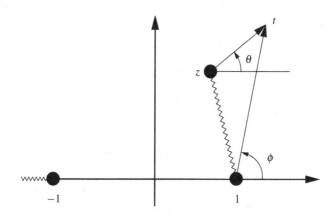

Fig. 6.3 The branch cut between the branch points $t = 1$ and $t = z$, and the definition of the angles ϕ and θ.

where L is the line segment from $t = 1$ to $t = z$ evaluated on the right-hand side of the cut. The branch cut between $t = 1$ and $t = z$ is parameterized by $t = 1 - u(1 - z)$, $u \in [0, 1]$.

$$\oint_C \frac{(t^2 - 1)^v}{(t - z)^{v+1}} \, dt$$

$$= 2ie^{-i\pi(v+1)} \sin \pi(v+1) \int_0^1 \frac{(2 - u(1-z))^v (u(z-1))^v}{(1 - z - u(1-z))^{v+1}} (z-1) \, du$$

$$= 2i \sin \pi(v+1) \int_0^1 (2 - u(1-z))^v u^v (1-u)^{-v-1} \, du$$

Compare this expression with the representation of the hypergeometric function in Theorem 5.3. We get from (6.2)

$$P_v(z) = F\left(-v, v+1; 1; \frac{1-z}{2}\right)$$

$$= \frac{2^{-v}}{\Gamma(v+1)\Gamma(-v)} \int_0^1 u^v (1-u)^{-v-1} (2 - u(1-z))^v \, du$$

valid for all $-1 < \operatorname{Re} v < 0$ and all $\mathbb{C} \ni z \notin (-\infty, -1]$. Make use of (A.4), and we get

$$P_v(z) = 2^{-v} \frac{\sin \pi(v+1)}{\pi} \int_0^1 (2 - u(1-z))^v u^v (1-u)^{-v-1} \, du$$

or

$$P_v(z) = \frac{2^{-v}}{2\pi i} \oint_C \frac{(t^2 - 1)^v}{(t - z)^{v+1}} \, dt$$

which completes the proof. □

This integral representation of the Legendre function of the first kind is often seen as an alternative definition of these functions.

We now turn to the integral representation of the Legendre function of the second kind, and a similar representation to the one for the Legendre functions of the first kind in Theorem 6.1 can be obtained. We have

Theorem 6.2 (Schläfli). *The Legendre function of the second kind, $Q_v(z)$, has an integral representation*

$$Q_v(z) = \frac{2^{-v}}{4i \sin v\pi} \oint_C \frac{(t^2 - 1)^v}{(z - t)^{v+1}} \, dt$$

where the contour C encircles the points $t = \pm 1$ in an eight-shape figure excluding the point $t = z$, see Figure 6.4. It is assumed that the argument z is not on the branch cut between the points $t = \pm 1$, i.e., $z \notin [-1, 1]$. The branch is chosen such that $\arg(1 - t) = \arg(1 + t) = 0$ on the upper part of the branch cut.

For $\operatorname{Re} v > -1$, the integral representation is identical to

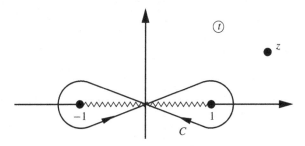

Fig. 6.4 The contour C in the complex t-plane in an eight-shape figure around $t = 1$ and $t = -1$ used in Theorem 6.2. The complex t-plane has a branch cut between the points $t = \pm 1$.

$$Q_v(z) = \frac{1}{2^{v+1}} \int_{-1}^{1} \frac{\left(1-t^2\right)^v}{(z-t)^{v+1}} \, dt$$

Proof. Note that the integrand resumes its original value after encircling the contour C. In fact, after encircling the point $t = -1$ in counter-clockwise direction, the integrand resumes its original value multiplied by a factor $e^{2iv\pi}$. Similarly, encircling the point $t = 1$ in clockwise direction, the integrand resumes its original value multiplied by a factor $e^{-2iv\pi}$. In total, the integrand resumes its original value after completing the full contour C.

To simplify the notation in the proof, we denote the right-hand side of the theorem by $I_v(z)$, and we have to prove that $I_v(z)$ is identical to the definition of (6.4). Deform the contour C to a line integral along the cut $[-1, 1]$ (one above the cut, and one below), and two circles around the points $t = \pm 1$, see Figure 6.5.

For $\operatorname{Re} v > -1$, the contributions from the two circles around $t = \pm 1$ vanish, and we obtain ($\mathbb{C} \ni z \notin [-1, 1]$)

$$I_v(z) = e^{iv\pi} \frac{2^{-v}}{4i \sin v\pi} \int_{-1}^{1} \frac{\left(1-t^2\right)^v}{(z-t)^{v+1}} \, dt - e^{-iv\pi} \frac{2^{-v}}{4i \sin v\pi} \int_{-1}^{1} \frac{\left(1-t^2\right)^v}{(z-t)^{v+1}} \, dt$$

$$= \frac{1}{2^{v+1}} \int_{-1}^{1} \frac{\left(1-t^2\right)^v}{(z-t)^{v+1}} \, dt$$

since $t - 1 = (1-t)e^{i\pi}$ and $t + 1 = 1 + t$ in the first integral (above the cut), and $t - 1 = (1-t)e^{i\pi}$ and $t + 1 = (1+t)e^{-2i\pi}$ in the second integral (below the cut).

As in the proof of Theorem 6.1, we prove the theorem by showing that the integral can be identified as the hypergeometric function for all $\operatorname{Re} v > -1$ and all $|z| > 1$.

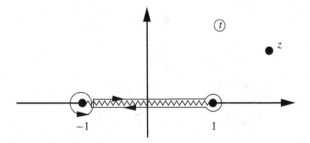

Fig. 6.5 The deformed contour in the complex t-plane around $t = 1$ and $t = -1$ with the branch cut between the points $t = \pm 1$.

$$I_\nu(z) = \frac{1}{(2z)^{\nu+1}} \int_{-1}^{1} (1-t^2)^\nu \left(1-\frac{t}{z}\right)^{-\nu-1} dt$$

$$= \frac{1}{(2z)^{\nu+1}} \int_{-1}^{1} (1-t^2)^\nu \sum_{k=0}^{\infty} (-1)^k \binom{-\nu-1}{k} \left(\frac{t}{z}\right)^k dt$$

$$= \frac{1}{(2z)^{\nu+1}} \int_{-1}^{1} (1-t^2)^\nu \sum_{k=0}^{\infty} \binom{-\nu-1}{2k} \left(\frac{t}{z}\right)^{2k} dt$$

since all odd powers in the integral give zero contribution. The binomial coefficient is rewritten as, see (A.14),

$$\binom{-\nu-1}{2k} = \frac{\Gamma(-\nu)}{\Gamma(-\nu-2k)(2k)!} = \frac{\Gamma(\nu+2k+1)}{\Gamma(\nu+1)(2k)!}$$

$$= \frac{2^\nu \Gamma(k+\nu/2+1/2)\Gamma(k+\nu/2+1)}{\Gamma(\nu+1)\Gamma(k+1/2)\Gamma(k+1)}$$

$$= \frac{2^\nu((\nu+1)/2,k)\Gamma(\nu/2+1/2)(\nu/2+1,k)\Gamma(\nu/2+1)}{\Gamma(\nu+1)\Gamma(k+1/2)k!}$$

where we also have used (A.4) and (A.5), i.e.,

$$\begin{cases} \Gamma(2k+1) = \dfrac{2^{2k}}{\sqrt{\pi}}\Gamma(k+1/2)\Gamma(k+1) \\[2mm] \Gamma(\nu+2k+1) = \dfrac{2^{\nu+2k}}{\sqrt{\pi}}\Gamma(k+\nu/2+1/2)\Gamma(k+\nu/2+1) \end{cases}$$

We get

$$I_\nu(z) = \frac{1}{2^{\nu+1}} \int_{-1}^{1} \frac{(1-t^2)^\nu}{(z-t)^{\nu+1}} dt$$

$$= \sum_{k=0}^{\infty} \frac{((\nu+1)/2,k)\Gamma(\nu/2+1/2)(\nu/2+1,k)\Gamma(\nu/2+1)}{z^{\nu+1}\Gamma(\nu+1)\Gamma(k+1/2)k!} \int_{0}^{1} (1-t^2)^\nu \left(\frac{t^2}{z^2}\right)^k dt$$

or

$$I_v(z) = \frac{1}{2}\frac{1}{z^{v+1}}\frac{\Gamma(v/2+1/2)\Gamma(v/2+1)}{\Gamma(v+3/2)}\sum_{k=0}^{\infty}\frac{((v+1)/2,k)(v/2+1,k)}{(v+3/2,k)}\frac{z^{-2k}}{k!}$$

$$= \frac{1}{2}\frac{1}{z^{v+1}}\frac{\Gamma(v/2+1/2)\Gamma(v/2+1)}{\Gamma(v+3/2)}F\left(1+\frac{v}{2},\frac{1+v}{2};v+\frac{3}{2};\frac{1}{z^2}\right)$$

where we used (A.16), i.e.,

$$\int_0^1 t^{2k}(1-t^2)^v\,dt = \frac{1}{2}\int_0^1 u^{k-1/2}(1-u)^v\,du = \frac{1}{2}\frac{\Gamma(k+1/2)\Gamma(v+1)}{\Gamma(v+k+3/2)}$$

$$= \frac{1}{2}\frac{\Gamma(k+1/2)\Gamma(v+1)}{(v+3/2,k)\Gamma(v+3/2)}$$

which is well defined as long as $\mathrm{Re}\,v > -1$. We finally get by the use of (6.4)

$$I_v(z) = \frac{1}{2^{v+1}}\int_{-1}^1\frac{(1-t^2)^v}{(z-t)^{v+1}}\,dt = \frac{2^v\Gamma(v/2+1/2)\Gamma(v/2+1)}{\sqrt{\pi}\Gamma(v+1)}Q_v(z) = Q_v(z)$$

since by (A.5)

$$\Gamma(v+1) = \frac{2^v}{\sqrt{\pi}}\Gamma(v/2+1/2)\Gamma(v/2+1)$$

and the proof is completed for $\mathrm{Re}\,v > -1$ and all $|z| > 1$. We thus have

$$Q_v(z) = \frac{2^{-v}}{4\mathrm{i}\sin v\pi}\oint_C\frac{(t^2-1)^v}{(z-t)^{v+1}}\,dt,\quad \mathrm{Re}\,v > -1 \text{ and } |z| > 1$$

By analytic continuation of both sides in this expression, the result holds in the full domain of the parameter v and the variable $\mathbb{C}\ni z\notin[-1,1]$. □

We illustrate the use of Theorem 6.2 by a more elaborate example.

Example 6.1. The Legendre function of the second kind, $Q_v(z)$, $z\in\mathbb{C}$, has an integral representation

$$Q_v(z) = \frac{1}{\sqrt{2}}\left\{\int_0^\pi\frac{\cos\left(v+\frac{1}{2}\right)t\,dt}{(z-\cos t)^{1/2}} - \cos v\pi\int_0^\infty\frac{e^{-\left(v+\frac{1}{2}\right)t}\,dt}{(z+\cosh t)^{1/2}}\right\} \tag{6.5}$$

where $\mathrm{Re}\,v > -1$, and where the principal branch of the square root is assumed, i.e., the square root (principal square root) is defined by a cut along the negative real axis, thus leading to

$$z^{1/2} = \sqrt{r}e^{\mathrm{i}\phi/2}\quad\text{if } z = re^{\mathrm{i}\phi},\quad \phi\in(-\pi,\pi]$$

In this example, we prove this representation.

For the parameter domain $\operatorname{Re} v > -1$, the Legendre function of the second kind, $Q_v(z)$, has an integral representation, see Theorem 6.2,

$$Q_v(z) = \frac{1}{2^{v+1}} \int_{-1}^{1} \frac{\left(1-t^2\right)^v}{(z-t)^{v+1}} \, dt$$

The representation in (6.5) is then proved by a change of variable, followed by a deformation of the contour.

To proceed, make a change of variables defined by

$$t \to u(t) = \frac{1-t^2}{2(z-t)} \tag{6.6}$$

The point $t = z$ is mapped to infinity in the u-plane, and $t = \pm 1$ are both mapped to the origin in the u-plane. The inverse is double-valued

$$t = u \pm \left(u^2 + 1 - 2zu\right)^{1/2} = u \pm (u_- - u)^{1/2} (u_+ - u)^{1/2}$$

The different signs refer to the two different Riemann sheets connected by a cut along a straight line connecting the two branch points $u_\pm$ in the u-plane. The points $u_\pm$ are explicitly given by

$$u_\pm = z \pm \left(z^2 - 1\right)^{1/2} = \frac{1}{2}\left((z+1)^{1/2} \pm (z-1)^{1/2}\right)^2 = |u_\pm| e^{i\theta_\pm}$$

The branch points satisfy

$$\begin{cases} u_+ + u_- = 2z \\ u_+ u_- = 1 \end{cases} \implies \begin{cases} |u_+| = 1/|u_-| \\ \theta_+ = -\theta_- \end{cases}$$

The image of the interval $t \in [-1, 1]$ under the transformation (6.6) is denoted C. The details of the properties of the contour C are left as an exercise to the reader, see Problem 6.3. The contour C starts at the origin in the u-plane, corresponding to $t = -1$, on the branch defined by

$$t = u - (u_- - u)^{1/2} (u_+ - u)^{1/2}$$

since

$$(u_-)^{1/2} (u_+)^{1/2} = e^{-i\theta/2} e^{i\theta/2} = 1$$

assuming the principal branch of the square root. The contour encircles the point u_- in a clockwise (counter-clockwise) direction if $\operatorname{Im} z > 0$ ($\operatorname{Im} z < 0$) and returns to the origin on the second Riemann sheet, see Figure 6.6. Moreover, in Problem 6.3 it is shown that u_- lies inside or on the unit circle, i.e., $|u_-| \leq 1$ and that u_+ satisfies $|u_+| \geq 1$.

Since the derivative of the transformation on the relevant Riemann sheet is

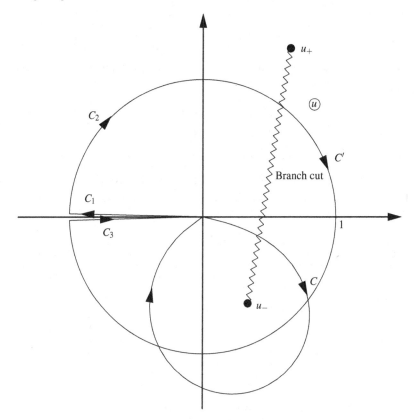

Fig. 6.6 The contour C and the deformed contour C' in the complex u-plane together with the branch cut connecting the branch points $u_\pm$ used in Example 6.1. In this particular example $z = 0.5 + i0.3$.

$$\frac{dt}{du} = \frac{z - t}{(u_- - u)^{1/2}(u_+ - u)^{1/2}}$$

the integral becomes

$$Q_\nu(z) = \frac{1}{2}\int_C \frac{u^\nu\, du}{(u_- - u)^{1/2}(u_+ - u)^{1/2}}$$

The singular points of the integrand are the two branch points $u_\pm$ and the origin.

The final step is to deform the contour C into the contour C' in Figure 6.6. In fact, any contour that starts at the origin and encircles the branch point u_- in a clockwise direction and returns to the origin on the second Riemann sheet gives the same value of the integral. Notice that no singular points, $u_\pm$, are crossed in this deformation, and the Legendre function of the second kind can be written as

$$Q_v(z) = \frac{1}{2} \int_{C'} \frac{u^v \, du}{(u_- - u)^{1/2} (u_+ - u)^{1/2}}$$

The contour C' is now parameterized, taking appropriate notice to the phase of the argument of the variable u

$$u = \begin{cases} e^{-w+i\pi}, & w \in (\infty, 0] & (C_1) \\ e^{i\phi}, & \phi \in [\pi, -\pi] & (C_2) \\ e^{-w-i\pi}, & w \in [0, \infty) & (C_3) \end{cases}$$

Notice that the value of the square root in the denominator along C_3 is identical to the corresponding values along C_1 apart from a sign difference. The parametric evaluation of this integral is

$$Q_v(z) = \frac{1}{2} \left(\int_0^\infty \frac{e^{-vw+iv\pi} e^{-w+i\pi} \, dw}{(e^{-2w} + 1 + 2ze^{-w})^{1/2}} + \int_0^\infty \frac{e^{-vw-iv\pi} e^{-w-i\pi} \, dw}{(e^{-2w} + 1 + 2ze^{-w})^{1/2}} \right)$$

$$- \frac{1}{2} \int_{-\pi}^\pi \frac{e^{iv\phi} i e^{i\phi} \, d\phi}{(u_- - e^{i\phi})^{1/2} (u_+ - e^{i\phi})^{1/2}}$$

$$= \frac{1}{2} \int_{-\pi}^\pi \frac{e^{i(v+1)\phi} \, d\phi}{\sqrt{2} e^{i\phi/2} (z - \cos\phi)^{1/2}} - \cos v\pi \int_0^\infty \frac{e^{-(v+1)w} \, dw}{(e^{-2w} + 1 + 2ze^{-w})^{1/2}}$$

since

$$\left(u_- - e^{-w\pm i\pi} \right)^{1/2} \left(u_+ - e^{-w\pm i\pi} \right)^{1/2} = \pm \left(e^{-2w} + 1 + 2ze^{-w} \right)^{1/2}, \quad \text{on } C_1/C_3$$

depending on the Riemann surface, and, moreover

$$\left(u_- - e^{i\phi} \right)^{1/2} \left(u_+ - e^{i\phi} \right)^{1/2} = -i\sqrt{2} e^{i\phi/2} (z - \cos\phi)^{1/2}; \quad \text{on } C_2$$

where the phase $-i$ had to be chosen so that the square root starts on the correct Riemann sheet at $\phi = \pi$. We simplify the expression to

$$Q_v(z) = \frac{1}{\sqrt{2}} \left(\int_0^\pi \frac{\cos(v+1/2)\phi}{(z - \cos\phi)^{1/2}} \, d\phi - \cos v\pi \int_0^\infty \frac{e^{-(v+1/2)w}}{(z + \cosh w)^{1/2}} \, dw \right)$$

where the principal branch of the square root is assumed. This completes the example. ∎

6.3 Associated Legendre functions

The associated Legendre functions are frequently used in applications, and they come in two families with slightly different definitions, i.e., $\{P_v^m(x), Q_v^m(x)\}$, and $\{\mathscr{P}_v^m(z), \mathscr{Q}_v^m(z)\}$, respectively. They all satisfy the differential equation

$$(z^2 - 1)u''(z) + 2zu'(z) - \left(v(v+1) + \frac{m^2}{z^2 - 1}\right)u(z) = 0$$

The parameter m can in general be an arbitrary complex constant, but in this section we restrict m to take integer values, which is the situation encountered in most applications. The differential equation has regular singular points at $z = \pm 1$ and at infinity.

The two families of the associated Legendre functions can be constructed from the Legendre functions of the first and second kind, $P_v(x)$, and $Q_v(x)$, respectively. To see this, first extract a factor $h(z) = (z^2 - 1)^{m/2}$ from the solution. Write $u(z) = h(z)v(z)$, and let $u(z)$ solve the differential equation above. By the use of Theorem 2.1, the function $v(z)$ then satisfies

$$(z^2 - 1)v''(z) + 2z(m+1)v'(z) + (m(m+1) - v(v+1))v(z) = 0$$

In order to compare with the hypergeometric differential equation, let $z = g(t) = 2t - 1$, and use Theorem 2.1 again to conclude that $w(t) = v(g(t))$ satisfies

$$t(t-1)w''(t) + (2t-1)(m+1)w'(t) + \underbrace{(m(m+1) - v(v+1))}_{(m+v+1)(m-v)}w(t) = 0$$

which is the hypergeometric differential equation with indices $\alpha = m - v$, $\beta = m + v + 1$, and $\gamma = m + 1$, and, as always, the order of the two first indices is immaterial. Repeated use of Lemma 4.1 on page 50 then shows that all solutions $w(t)$ can be written as the mth order derivatives of the solutions to

$$t(t-1)w''(t) + (2t-1)w'(t) - v(v+1)w(t) = 0$$

which is the Legendre differential equation. This motivates the following two definitions of the associated Legendre functions for non-negative integer m. The first one is

$$\begin{cases} P_v^m(x) = (1-x^2)^{\frac{m}{2}} \dfrac{d^m}{dx^m} P_v(x) \\[2mm] Q_v^m(x) = (1-x^2)^{\frac{m}{2}} \dfrac{d^m}{dx^m} Q_v(x) \end{cases} \qquad x \in [-1,1]$$

and the second is

$$\begin{cases} \mathscr{P}_v^m(z) = (z^2 - 1)^{\frac{m}{2}} \dfrac{d^m}{dz^m} P_v(z) \\[3mm] \mathscr{Q}_v^m(z) = (z^2 - 1)^{\frac{m}{2}} \dfrac{d^m}{dz^m} Q_v(z) \end{cases} \qquad z \in \mathbb{C}$$

The two families $\{P_v^m(x), Q_v^m(x)\}$ and $\{\mathscr{P}_v^m(z), \mathscr{Q}_v^m(z)\}$ are related to each other, and they differ in the domain of definition and a common phase factor. The first family is only defined in the interval $x \in [-1, 1]$, i.e., on the line between the two regular singular points $z = \pm 1$. The second family is defined everywhere in the complex plane except on the line between the two regular singular points $z = \pm 1$.

From the definition of $P_v(z)$ in terms of the hypergeometric function in (6.2) and the result in Problem 5.2, we get

$$\begin{aligned} \mathscr{P}_v^m(z) &= (z^2 - 1)^{\frac{m}{2}} \frac{d^m}{dz^m} P_v(z) \\ &= \frac{(-1)^m}{2^m} \frac{(-v, m)(v+1, m)}{(1, m)} (z^2 - 1)^{\frac{m}{2}} F\left(m - v, m + v + 1; m + 1; \frac{1-z}{2}\right) \\ &= \frac{(-1)^m}{2^m} \frac{\Gamma(m-v)\Gamma(v+m+1)}{m!\Gamma(-v)\Gamma(v+1)} (z^2 - 1)^{\frac{m}{2}} F\left(m - v, m + v + 1; m + 1; \frac{1-z}{2}\right) \end{aligned}$$

Use (A.4), and we obtain

$$\mathscr{P}_v^m(z) = \frac{\Gamma(v+m+1)}{2^m m! \Gamma(v-m+1)} (z^2 - 1)^{\frac{m}{2}} F\left(m - v, m + v + 1; m + 1; \frac{1-z}{2}\right) \quad (6.7)$$

which is an alternative way of defining the associated Legendre functions of the first kind.

Example 6.2. The Helmholtz equation does not separate in the toroidal coordinate system. However, the Laplace equation does, and the differential equations generated by separation of variables are related to the Legendre functions.

The toroidal coordinates (ξ, η, ϕ) are defined as

$$(x_1, x_2, x_3) = \frac{a}{\cosh \eta - \cos \xi} (\sinh \eta \cos \phi, \sinh \eta \sin \phi, \sin \xi)$$

where the domains of the coordinates are

$$\xi \in [0, 2\pi), \qquad \eta \in [0, \infty), \qquad \phi \in [0, 2\pi)$$

The Laplace equation, $\nabla^2 \psi(\xi, \eta, \phi) = 0$, is separable, if we extract a common factor $\sqrt{\cosh \eta - \cos \xi}$, i.e.,

$$\psi(\xi, \eta, \phi) = \sqrt{\cosh \eta - \cos \xi} \, F(\xi, \eta, \phi)$$

Then $F(\xi, \eta, \phi)$ satisfies

$$\frac{\partial^2 F}{\partial \xi^2} + \frac{1}{\sinh \eta}\frac{\partial}{\partial \eta}\left(\sinh \eta \frac{\partial F}{\partial \eta}\right) + \frac{1}{\sinh^2 \eta}\frac{\partial^2 F}{\partial \phi^2} + \frac{1}{4}F = 0$$

The appropriate systems in the ξ and the ϕ variables are the trigonometric functions, i.e.,

$$\begin{pmatrix}\cos n\xi\\ \sin n\xi\end{pmatrix} \qquad \begin{pmatrix}\cos m\phi\\ \sin m\phi\end{pmatrix}, \qquad m,n \text{ non-negative integers}$$

which implies

$$\frac{1}{\sinh \eta}\frac{d}{d\eta}\left(\sinh \eta \frac{df(\eta)}{d\eta}\right) - \frac{m^2}{\sinh^2 \eta}f(\eta) - \left(n^2 - \frac{1}{4}\right)f(\eta) = 0$$

Introduce the independent variable $z = \cosh \eta$, and the equation is transformed into

$$\frac{d}{dz}\left((z^2 - 1)\frac{df(z)}{dz}\right) - \frac{m^2}{z^2 - 1}f(z) - (n - 1/2)(n + 1/2)f(z) = 0$$

The solutions of this differential equation are most conveniently expressed by the use of the definition in (6.7), extended to non-integer values of m. The solutions are then $\mathscr{P}^m_{n-\frac{1}{2}}(\cosh \eta)$ or $\mathscr{Q}^m_{n-\frac{1}{2}}(\cosh \eta)$ — the toroidal functions or the ring functions.
∎

Problems

6.1. Prove the representation of the Legendre functions of the first kind, due to Laplace,[3] for $\mathrm{Re}\, z > 0$

$$P_\nu(z) = \frac{1}{\pi}\int_0^\pi \left(z + (z^2 - 1)^{1/2}\cos\phi\right)^\nu d\phi, \qquad \mathrm{Re}\, z > 0$$

where the branch of the square root $(z^2 - 1)^{1/2}$ is real positive when z is real and > 1.
Hint: Use the following parameter representation of the contour C in Schläfli's integral:

$$t = z + (z^2 - 1)^{1/2}e^{i\phi}, \qquad \phi \in (-\pi, \pi]$$

6.2. [†]Prove the representation of the Legendre functions of the first kind, due to Dirichlet[4] and Mehler,[5] for $\theta \in (0, \pi)$

---

[3] Pierre-Simon Laplace (1749–1827), French mathematician.
[4] Peter Gustav Lejeune Dirichlet (1805–1859), German mathematician.
[5] Ferdinand Mehler, German mathematician.

$$P_\nu(\cos\theta) = \frac{\sqrt{2}}{\pi} \int_0^\theta \frac{\cos\left(\nu+\frac{1}{2}\right)\phi}{(\cos\phi - \cos\theta)^{1/2}}\,d\phi, \quad \theta \in (0,\pi)$$

where the branch of the square root is real positive when the argument is real and positive.

Hint: Use the following parameter representation of the contour C in Schläfli's integral:

$$t = 1 + 2\sin\frac{\theta}{2}\,e^{i\phi}, \quad \phi \in (-(\pi+\theta)/2, (3\pi-\theta)/2]$$

6.3. Define the transformation from the complex t-plane to the u-plane by

$$t \to u(t) = \frac{1-t^2}{2(z-t)}$$

where $\mathbb{C} \ni z \notin [-1,1]$. Show that the interval $t \in [-1,1]$ is mapped to a contour C, see Figure 6.6, starting at the origin in the u-plane, encircling the point $u_- = z - (z^2-1)^{1/2}$ in a clockwise direction, and returning to the origin. Show that the point u_- lies inside or on the unit circle in the complex u-plane.

6.4. [†]Show the relation ($m \in \mathbb{N}$) [15, p. 182]

$$\frac{\sinh^m\eta}{(\cosh\eta - \cos\xi)^{m+\frac{1}{2}}} = \frac{(-1)^m\sqrt{2}}{\sqrt{\pi}\,\Gamma\left(m+\frac{1}{2}\right)} \sum_{n=0}^{\infty} \varepsilon_n \mathcal{Q}_{n-\frac{1}{2}}^m(\cosh\eta)\cos n\xi$$

where the Neumann[6] symbol is $\varepsilon_n = 2 - \delta_{n,0}$.

6.5. Using the result in Problem 6.4, solve the integral ($m,n \in \mathbb{N}$)

$$\int_0^\pi \frac{\cos n\xi}{(\cosh\eta_0 - \cos\xi)^{m+\frac{1}{2}}}\,d\xi$$

---

[6] Carl Gottfried Neumann (1832–1925), German mathematician.

Chapter 7
Confluent hypergeometric functions

The details of the hypergeometric function were developed in Chapter 5. We used the notion $F(\alpha,\beta;\gamma;z)$ to denote this function. A more complete notation on the hypergeometric function is $_2F_1(\alpha,\beta;\gamma;z)$, which, in this chapter, is used in parallel with the shorter one.

What happens if two of the regular singular points at $z = 0, 1, \infty$ in the hypergeometric differential equation coalesce is the topic of this chapter. The way we let this happen is that we scale the singular point at $z = 1$ and push it to infinity, where already one of the regular singular points resides. In order to get something new and meaningful out from this limit process, we also scale the roots, α and β, of the indicial equation. There are two ways in which this scaling (or confluence) can occur, and we treat these cases in two separate sections. In the first scaling, see Section 7.1, one of the roots of the indicial equation approaches infinity, while, in the second scaling, both of the roots approach infinity simultaneously, see Section 7.2.

7.1 Confluent hypergeometric functions — first kind

Of interest in this section is the confluence of the first kind. More explicitly, if the three regular singular points are located at 0, b, and ∞, and one of the roots of the indicial equation, corresponding to the point at infinity, is $\beta = kb$, we let $b \to \infty$, where k is an independent complex constant in this limit process. In terms of Riemann's P symbol this is

$$
\lim_{b\to\infty} P \left\{ \begin{array}{ccc} 0 & 1 & \infty \\ 0 & 0 & \alpha \quad z/b \\ 1-\gamma & \gamma-\alpha-kb & kb \end{array} \right\} = \lim_{b\to\infty} P \left\{ \begin{array}{ccc} 0 & b & \infty \\ 0 & 0 & \alpha \quad z \\ 1-\gamma & \gamma-\alpha-kb & kb \end{array} \right\}
$$

We now study what happens to the hypergeometric series in this limit process

G. Kristensson, *Second Order Differential Equations: Special Functions and Their Classification*, DOI 10.1007/978-1-4419-7020-6_7,
© Springer Science+Business Media, LLC 2010

$$_2F_1(\alpha, kb; \gamma; z/b) = \sum_{n=0}^{\infty} \frac{(\alpha, n)(kb, n)}{(1, n)(\gamma, n)} \left(\frac{z}{b}\right)^n$$

The factor

$$\frac{(kb, n)}{b^n} = k(k + 1/b) \cdot \ldots \cdot (k + (n - 1)/b) \to k^n \text{ as } b \to \infty$$

Therefore, we get for z inside the domain of convergence

$$\lim_{b \to \infty} {}_2F_1(\alpha, kb; \gamma; z/b) = \sum_{n=0}^{\infty} \frac{(\alpha, n)}{(1, n)(\gamma, n)} (kz)^n$$

This power series converges everywhere in the finite complex plane,[1] i.e., $|z| < \infty$. We introduce the notation, also called Kummer's function or the M-function

$$_1F_1(\alpha; \gamma; z) = \sum_{n=0}^{\infty} \frac{(\alpha, n)}{(\gamma, n)} \frac{z^n}{n!} \tag{7.1}$$

satisfying

$$_1F_1(\alpha; \gamma; 0) = 1$$

and Kummer's differential equation for $u(z) = {}_1F_1(\alpha; \gamma; z)$ is

$$zu''(z) + (\gamma - z)u'(z) - \alpha u(z) = 0$$

which is obtained by taking the limit $b \to \infty$ in the scaled version ($\beta = b, z \to z/b$) of the hypergeometric equation (4.3). We observe that this differential equation has one regular singular point at the origin, and one irregular singular point at infinity. Therefore, the confluence of the first kind has created an irregular singular point out of the two regular singular points at $z = b \to \infty$ and infinity.

The indicial equation of Kummer's differential equation, at the regular singular point at the origin, has roots 0 and $1 - \gamma$, respectively, and the second solution to the differential equation is

$$v(z) = z^{1-\gamma} f(z)$$

where, using Theorem 2.1 on page 4,

$$zf''(z) + (2 - \gamma - z)f'(z) - (\alpha - \gamma + 1)u(z) = 0$$

This differential equation has a solution, which is well behaved at the origin, viz.,

$$f(z) = {}_1F_1(\alpha - \gamma + 1; 2 - \gamma; z)$$

---

[1] Compare with the convergence of the exponential function

$$e^z = \sum_{n=0}^{\infty} \frac{z^n}{(1, n)}$$

This observation implies that

$$v(z) = z^{1-\gamma}{}_1F_1(\alpha - \gamma + 1; 2 - \gamma; z), \quad \gamma \text{ not an integer} \tag{7.2}$$

is a second solution to the differential equation.

7.1.1 Bessel functions

The Bessel[2] functions, $J_v(z)$, where v is in general a complex parameter, play an important role in mathematical physics, and they are related to the confluent hypergeometric function, ${}_1F_1(\gamma; z)$. The aim of this section is to show this relationship in more detail.

The Bessel function $J_v(z)$ satisfies the Bessel differential equation

$$z^2 u''(z) + z u'(z) + (z^2 - v^2) u(z) = 0 \tag{7.3}$$

and the Bessel function $J_v(z)$ behaves near the origin as

$$J_v(z) \sim \frac{z^v}{2^v \Gamma(v+1)} \tag{7.4}$$

The Bessel differential equation and the confluent hypergeometric equation

$$z u''(z) + (\gamma - z) u'(z) - \alpha u(z) = 0$$

are related. To see this, use Theorem 2.1 on page 4 and $g(t) = 2it$. The confluent hypergeometric equation is then transformed into

$$t w''(t) + (\gamma - 2it) w'(t) - 2i\alpha w(t) = 0$$

Another application of Theorem 2.1 with $w(t) = t^{-v} e^{it} u(t)$ implies

$$t^2 u''(t) + (\gamma - 2v) t u'(t) + \left((v - it)^2 + v + (\gamma - 2it)(-v + it) - 2i\alpha t \right) u(t) = 0$$

From these transformations, we see that by choosing $\alpha = v + 1/2$ and $\gamma = 2v + 1$, we get

$$t^2 u''(t) + t u'(t) + (t^2 - v^2) u(t) = 0$$

which is identical to the Bessel differential equation. This leads to the identification or definition of the Bessel function, $J_v(z)$, in terms of the confluent hypergeometric function, ${}_1F_1(\gamma; z)$,

$$J_v(z) = \frac{z^v e^{-iz} {}_1F_1(v + 1/2; 2v + 1; 2iz)}{2^v \Gamma(v+1)} \tag{7.5}$$

---

[2] Friedrich Bessel (1784–1846), German mathematician.

where the multiplicative constant is adjusted to the correct behavior as $z \to 0$, see (7.4).

The properties of the Bessel function are treated in detail by Watson [30], and the reader is encouraged to study that excellent text in order to find out more about the many different integral representations and additional properties of the Bessel functions. We restrict ourselves to give an example of integral representations of Bessel functions in Example 7.1 below.

7.1.2 Integral representations

From the integral representations in Chapter 5, we inherit integral representations for the confluent hypergeometric function $_1F_1(\alpha; \gamma; z)$. We start with the one developed in Section 5.5.

Theorem 7.1. *Kummer's function is represented as an integral*

$$_1F_1(\alpha; \gamma; z) = \frac{\Gamma(\gamma)}{\Gamma(\alpha)\Gamma(\gamma - \alpha)} \int_0^1 u^{\alpha-1}(1-u)^{\gamma-\alpha-1} e^{uz} \, du$$

valid for all $\operatorname{Re} \gamma > \operatorname{Re} \alpha > 0$ *and all* $z \in \mathbb{C}$.

Proof. We start with Theorem 5.3 where α and β have been interchanged (has no effect on the hypergeometric function)

$$F(\alpha, \beta; \gamma; z/\beta) = \frac{\Gamma(\gamma)}{\Gamma(\alpha)\Gamma(\gamma - \alpha)} \int_0^1 u^{\alpha-1}(1-u)^{\gamma-\alpha-1}(1-uz/\beta)^{-\beta} \, du$$

valid for all $\operatorname{Re} \gamma > \operatorname{Re} \alpha > 0$ and all $\mathbb{C} \ni z/b \notin [1, \infty)$. The result of the theorem is then immediately obtained by taking the limit $\beta \to \infty$ inside the integral using the following representation of the exponential:

$$e^z = \lim_{b \to \infty} \left(1 + \frac{z}{b}\right)^b \tag{7.6}$$

The restriction on the variable z can be dropped, since the integral converges for all $z \in \mathbb{C}$. $\square$

Example 7.1. The Bessel function $J_\nu(z)$ was defined in (7.5).

$$J_\nu(z) = \left(\frac{z}{2}\right)^\nu \frac{e^{-iz} {}_1F_1(\nu + 1/2; 2\nu + 1; 2iz)}{\Gamma(\nu + 1)}$$

The integral representation in Theorem 7.1 is used to express the Bessel function as an integral. We obtain, $\operatorname{Re} \nu > -1/2$,

$$J_\nu(z) = \left(\frac{z}{2}\right)^\nu \frac{e^{-iz}}{\Gamma(\nu + 1)} \frac{\Gamma(2\nu + 1)}{(\Gamma(\nu + 1/2))^2} \int_0^1 u^{\nu-1/2}(1-u)^{\nu-1/2} e^{2iuz} \, du$$

The gamma functions are simplified using (A.5) with $z = v + 1/2$, i.e.,

$$\Gamma(2v+1) = \frac{2^{2v}}{\sqrt{\pi}}\Gamma(v+1/2)\Gamma(v+1)$$

Also introduce the change of variable $u = (1-t)/2$. We obtain

$$J_v(z) = \left(\frac{z}{2}\right)^v \frac{1}{\sqrt{\pi}\,\Gamma(v+1/2)} \int_{-1}^{1}(1-t^2)^{v-1/2}e^{-itz}\,dt$$

Another change of variable $t = -\cos\theta$ puts the integral representation into

$$J_v(z) = \left(\frac{z}{2}\right)^v \frac{1}{\sqrt{\pi}\,\Gamma(v+1/2)} \int_{0}^{\pi} \sin^{2v}\theta\,e^{iz\cos\theta}\,d\theta, \qquad \mathrm{Re}\,v > -1/2$$

This integral representation originates from Siméon-Denis Poisson (1781–1840), who was a French mathematician. ∎

There is also a parallel to Barnes' integral representation in Theorem 5.4 on page 80 for Kummer's function.

Theorem 7.2. *If α or γ are not zero or a negative integer, Kummer's function can be represented as a contour integral*

$$_1F_1(\alpha;\gamma;z) = \frac{1}{2\pi i}\int_E \frac{\Gamma(\gamma)\Gamma(\alpha+s)}{\Gamma(\alpha)\Gamma(\gamma+s)}\Gamma(-s)(-z)^s\,ds, \quad |\arg(-z)| < \pi/2$$

where E is a contour in the complex s-plane, starting at $-i\infty$ and ending at $i\infty$, such that all poles to $\Gamma(\alpha+s)$ (i.e., $s = -\alpha, -\alpha-1, -\alpha-2,\ldots$) lie to the left and all poles to $\Gamma(-s)$ (i.e., $s \in \mathbb{N}$) lie to the right of E, see Figure 7.1.

Proof. The proof follows the proof of Barnes' theorem closely. We rewrite the gamma functions in the integrand.

$$\frac{\Gamma(\alpha+s)}{\Gamma(\gamma+s)}\Gamma(-s) = \frac{\Gamma(\alpha+s)}{\Gamma(s)}\frac{\Gamma(s)}{\Gamma(\gamma+s)}\Gamma(-s)$$

Lemma A.3 and Lemma A.4 on page 170 and page 172, respectively, show that for each $\varepsilon \in (0,1/2)$ we can estimate the gamma functions in the integrand as

$$\left|\frac{\Gamma(\alpha+s)}{\Gamma(\gamma+s)}\Gamma(-s)\right| \leq Ce^{\mathrm{Re}(\alpha-\gamma-s-1/2)\ln|s|+\mathrm{Re}\,s+\mathrm{Im}\,s\arg(-s)}$$

when $s \in \{s \in \mathbb{C} : |s+\alpha+n| \geq \varepsilon, |s-n| \geq \varepsilon, \forall n \in \mathbb{N}\}$. Note that this factor has no singularity at $s = -\gamma-n$ or $s = -n$, $n \in \mathbb{N}$. We also estimate

$$|(-z)^s| = \left||z|^s e^{is\arg(-z)}\right| = e^{\mathrm{Re}\,s\ln|z|-\mathrm{Im}\,s\arg(-z)}$$

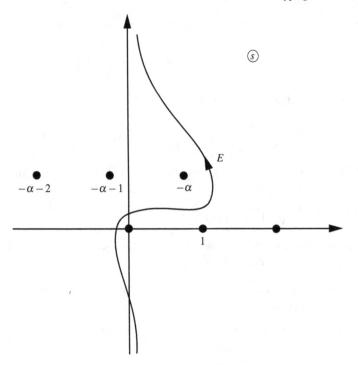

Fig. 7.1 The contour of integration in Theorem 7.2.

The integrand can now be estimated. We get

$$\left|\frac{\Gamma(\alpha+s)}{\Gamma(\gamma+s)}\Gamma(-s)(-z)^s\right| \le Ce^{\text{Re}(\alpha-\gamma-s-1/2)\ln|s|+\text{Re}s(1+\ln|z|)+\text{Im}s\arg(-s)-\text{Im}s\arg(-z)}$$

Using

$$\arg(-s) = \arg(s) - \begin{cases} \pi, & \text{Im}\,s \ge 0 \\ -\pi, & \text{Im}\,s \le 0 \end{cases}$$

we get

$$\left|\frac{\Gamma(\alpha+s)}{\Gamma(\gamma+s)}\Gamma(-s)(-z)^s\right| \le Ce^{\text{Re}(\alpha-\gamma-s-1/2)\ln|s|+\text{Re}s(1+\ln|z|)+|\text{Im}s|(|\arg(s)|-\pi+|\arg(-z)|)}$$

$$(7.7)$$

For large values of s along the imaginary axis, the integrand is dominated by

$$\left|\frac{\Gamma(\alpha+s)}{\Gamma(\gamma+s)}\Gamma(-s)(-z)^s\right| \le Ce^{\text{Re}(\alpha-\gamma-s-1/2)\ln|s|+\text{Re}s(1+\ln|z|)+|\text{Im}s|(|\arg(-z)|-\pi/2)}$$

$$\le Ce^{\text{Re}(\alpha-\gamma-s-1/2)\ln|s|+\text{Re}s(1+\ln|z|)-|\text{Im}s|\delta}$$

where we have assumed $|\arg(-z)| < \pi/2 - \delta$, $\delta > 0$. The integrand is then exponentially small as $s \to \pm i\infty$, and the integral is convergent, provided $|\arg(-z)| < \pi/2$. This concludes the first part of the proof.

We next show that the integral in the theorem is identical to Kummer's function in (7.1). As in the proof of Lemma 5.5, we close the contour by a semi-circle C in the right-hand side of the s-plane. The poles at $s \in \mathbb{N}$ are avoided by parameterizing the contour C as

$$C: s = (N + 1/2) e^{i\phi} = (N + 1/2)(\cos\phi + i\sin\phi), \qquad \phi \in [-\pi/2, \pi/2]$$

and let $N \to \infty$ through the integer numbers. The integrand on C is bounded by the use of the estimates above, see (7.7),

$$\left| \frac{\Gamma(\alpha + s)}{\Gamma(\gamma + s)} \Gamma(-s)(-z)^s \right| \leq Ce^{\mathrm{Re}(\alpha - \gamma - 1/2)\ln(N+1/2) + (N+1/2)\cos\phi(1 + \ln|z| - \ln(N+1/2))}$$

$$\times e^{(N+1/2)|\sin\phi|(|\phi| - \pi + |\arg(-z)|)}$$

or, provided $|\arg(-z)| < \pi/2 - \delta$, $\delta > 0$,

$$\left| \frac{\Gamma(\alpha + s)}{\Gamma(\gamma + s)} \Gamma(-s)(-z)^s \right| \leq Ce^{\mathrm{Re}(\alpha - \gamma - 1/2)\ln(N+1/2) + (N+1/2)\cos\phi(1 + \ln|z| - \ln(N+1/2))}$$

$$\times e^{(N+1/2)|\sin\phi|(|\phi| - \pi/2 - \delta)}$$

We see, using the same type of arguments as in Lemma 5.5,[3] that the contribution from the contour C vanishes as $N \to \infty$, and the residue theorem implies

$$\int_E \frac{\Gamma(\gamma)\Gamma(\alpha + s)}{\Gamma(\alpha)\Gamma(\gamma + s)} \Gamma(-s)(-z)^s \, ds = \int_{E-C} \frac{\Gamma(\gamma)\Gamma(\alpha + s)}{\Gamma(\alpha)\Gamma(\gamma + s)} \Gamma(-s)(-z)^s \, ds$$

$$= -2\pi i \sum_{n=0}^{\infty} \operatorname*{Res}_{s=n} \frac{\Gamma(\gamma)\Gamma(\alpha + s)}{\Gamma(\alpha)\Gamma(\gamma + s)} \Gamma(-s)(-z)^s$$

The residues of $\Gamma(-s)$, see (5.15) on page 84, then imply

$$\int_E \frac{\Gamma(\gamma)\Gamma(\alpha + s)}{\Gamma(\alpha)\Gamma(\gamma + s)} \Gamma(-s)(-z)^s \, ds$$

$$= 2\pi i \sum_{n=0}^{\infty} \frac{\Gamma(\gamma)\Gamma(\alpha + n)}{\Gamma(\alpha)\Gamma(\gamma + n)} \frac{(-1)^n}{n!} (-z)^n = 2\pi i \sum_{n=0}^{\infty} \frac{(\alpha, n)}{(\gamma, n)} \frac{z^n}{n!}$$

and the theorem is proved. □

---

[3] In the interval $|\phi| < \pi/4$, the factor $\exp\{-(N + 1/2)\cos\phi \ln(N + 1/2)\}$, which is exponentially small as $N \to \infty$, dominates, and in the interval $\pi/4 < |\phi| \leq \pi/2$, the factor $\exp\{N + 1/2)|\sin\phi|(|\phi| - \pi/2 - \delta\}$ is exponentially small as $N \to \infty$.

7.1.3 Laguerre polynomials

Polynomial solutions are obtained by letting $\alpha = -n$, $n \in \mathbb{N}$, in the confluent hypergeometric function $_1F_1(\alpha; \gamma; z)$. These polynomials are the Laguerre[4] polynomials $L_n^{(\alpha)}(x)$, which with suitable normalization are defined as

$$L_n^{(\alpha)}(x) = \binom{n+\alpha}{n} {}_1F_1(-n; \alpha+1; x) \tag{7.8}$$

The Laguerre polynomials, $L_n^{(\alpha)}(x)$, satisfy

$$zu''(z) + (\alpha+1-z)u'(z) + nu(z) = 0$$

The confluent hypergeometric function, $_1F_1(\alpha; \gamma; z)$, was obtained as a limit process from the hypergeometric function, $_2F_1(\alpha, \beta; \gamma; z)$, i.e., scaling of β and the location of the regular singular point at $z = 1$. The same procedure gives a relation between the Laguerre polynomials and the Jacobi polynomials defined in (5.18) on page 94. Since the normalization coefficients for both sets of polynomials are the same, compare (5.18) and (7.8), the result is

$$L_n^{(\alpha)}(x) = \lim_{b \to \infty} P_n^{(\alpha,b)}\left(1 - \frac{2x}{b}\right) \tag{7.9}$$

With this observation, Rodrigues' generalized function, the generating function, and the orthogonality relation of the Laguerre polynomials are obtained from the results in Section 5.8 on page 92.

The argument x in the Laguerre polynomials does usually take positive values, in contrast to the Jacobi polynomials where the arguments usually take values in the interval $[-1, 1]$, see, e.g., the orthogonality relation in Lemma 5.8.

We derive Rodrigues' generalized function for the Laguerre polynomials using the result in Section 5.8.3, and taking the appropriate limit $b \to \infty$ in the Jacobi polynomials, see (7.9). The limit, using (5.25), is

$$L_n^{(\alpha)}(x)$$
$$= \frac{(-1)^n}{2^n n!} \lim_{b \to \infty} \left(\frac{2x}{b}\right)^{-\alpha} \left(2 - \frac{2x}{b}\right)^{-b} \left(-\frac{b}{2}\right)^n \frac{d^n}{dx^n}\left\{\left(\frac{2x}{b}\right)^{\alpha+n}\left(2 - \frac{2x}{b}\right)^{b+n}\right\}$$
$$= \frac{1}{n!} \lim_{b \to \infty} x^{-\alpha}\left(1 - \frac{x}{b}\right)^{-b} \frac{d^n}{dx^n}\left\{x^{\alpha+n}\left(1 - \frac{x}{b}\right)^{b+n}\right\}$$

which reduces to, see (7.6),

$$L_n^{(\alpha)}(x) = \frac{1}{n!} x^{-\alpha} e^x \frac{d^n}{dx^n}\left\{x^{\alpha+n} e^{-x}\right\}$$

---

[4] Edmond Laguerre (1834–1886), French mathematician.

Similarly, the generating function for the Laguerre polynomials, $G^{(\alpha)}(x,t)$, is obtained from (5.24) on page 98 by a limit process

$$G^{(\alpha)}(x,t) = \sum_{n=0}^{\infty} L_n^{(\alpha)}(x)t^n = \sum_{n=0}^{\infty} \lim_{b\to\infty} P_n^{(\alpha,b)}\left(1 - \frac{2x}{b}\right)t^n$$

$$= \lim_{b\to\infty} \frac{2^{\alpha+b}\left((1-t)^2 + \frac{4xt}{b}\right)^{-1/2}}{\left(1-t+\sqrt{(1-t)^2 + \frac{4xt}{b}}\right)^{\alpha}\left(1+t+\sqrt{(1-t)^2 + \frac{4xt}{b}}\right)^{b}}$$

$$= \lim_{b\to\infty} \frac{2^b(1-t)^{-\alpha-1}}{\left(2 + \frac{2xt}{b(1-t)} + O(1/b^2)\right)^b} = (1-t)^{-\alpha-1}e^{-\frac{xt}{1-t}}$$

The final result then is

$$G^{(\alpha)}(x,t) = \sum_{n=0}^{\infty} L_n^{(\alpha)}(x)t^n = (1-t)^{-\alpha-1}\exp\left\{-\frac{xt}{1-t}\right\}$$

Finally, we use the generating function, $G^{(\alpha)}(x,t)$, to find the orthogonality properties of the Laguerre polynomials. The following double sum is appropriate:

$$\sum_{n=0}^{\infty}\sum_{m=0}^{\infty}\int_0^{\infty} e^{-x}x^{\alpha}L_n^{(\alpha)}(x)L_m^{(\alpha)}(x)t^n u^m\,dx = \int_0^{\infty} e^{-x}x^{\alpha}G^{(\alpha)}(x,t)G^{(\alpha)}(x,u)\,dx$$

$$= \int_0^{\infty} e^{-x}x^{\alpha}(1-t)^{-\alpha-1}\exp\left\{-\frac{xt}{1-t}\right\}(1-u)^{-\alpha-1}\exp\left\{-\frac{xu}{1-u}\right\}dx$$

$$= (1-u)^{-\alpha-1}(1-t)^{-\alpha-1}\int_0^{\infty}x^{\alpha}\exp\left\{-x\left(1+\frac{t}{1-t}+\frac{u}{1-u}\right)\right\}dx$$

Further calculations give

$$\sum_{n=0}^{\infty}\sum_{m=0}^{\infty}\int_0^{\infty} e^{-x}x^{\alpha}L_n^{(\alpha)}(x)L_m^{(\alpha)}(x)t^n u^m\,dx$$

$$= (1-u)^{-\alpha-1}(1-t)^{-\alpha-1}\Gamma(\alpha+1)\left(1+\frac{t}{1-t}+\frac{u}{1-u}\right)^{-\alpha-1}$$

$$= \Gamma(\alpha+1)(1-ut)^{-\alpha-1} = \Gamma(\alpha+1)\sum_{n=0}^{\infty}\binom{-1-\alpha}{n}(-ut)^n$$

$$= \Gamma(\alpha+n+1)\sum_{n=0}^{\infty}\frac{(ut)^n}{n!}$$

where we used the definition of the gamma function $\Gamma(z)$, (A.1), in Appendix A to obtain ($a > 0$)

$$\int_0^{\infty} e^{-ax}x^{\alpha}\,dx = a^{-\alpha-1}\int_0^{\infty} e^{-t}t^{\alpha}\,dt = a^{-\alpha-1}\Gamma(\alpha+1)$$

and by using (A.4)

$$\binom{-\beta}{n} = \frac{\Gamma(1-\beta)}{\Gamma(1-\beta-n)n!} = (-1)^n \frac{\Gamma(\beta+n)}{\Gamma(\beta)n!}$$

The coefficients in front of the powers in t and u on the left- and the right-hand side must be identical, leading to the orthogonality result for the Laguerre polynomials

$$\int_0^\infty e^{-x} x^\alpha L_n^{(\alpha)}(x) L_m^{(\alpha)}(x) \, dx = \frac{\Gamma(\alpha+n+1)}{n!} \delta_{n,m}$$

7.1.4 Hermite polynomials

We conclude this section by defining a set of polynomials —— the Hermite polynomials — as a special case of the Laguerre polynomials. Specifically, the Hermite[5] polynomials, $H_m(x)$, are defined in terms of the Laguerre polynomials by letting $\alpha = \pm 1/2$, viz.,

$$\begin{cases} H_{2m}(x) = (-1)^m 2^{2m} m! L_m^{(-1/2)}(x^2) \\ H_{2m+1}(x) = (-1)^m 2^{2m+1} m! x L_m^{(1/2)}(x^2) \end{cases} \quad x \in \mathbb{R}$$

The argument of the Hermite polynomials usually takes values on the whole real axis.

7.2 Confluent hypergeometric functions — second kind

The second kind of confluence in the hypergeometric function, $F(\alpha,\beta;\gamma;z) = {}_2F_1(\alpha,\beta;\gamma;z)$, where both roots of the indicial equation approach infinity together with one of the regular singular points, is now investigated. To this end, let $\alpha = k_1\sqrt{b}$ and $\beta = k_2\sqrt{b}$, and study what happens to the hypergeometric series as $b \to \infty$ with a scaled argument z/b. The hypergeometric series in this limit process is

$$\lim_{b\to\infty} {}_2F_1(k_1\sqrt{b}, k_2\sqrt{b}; \gamma; z/b) = \sum_{n=0}^\infty \frac{(k_1\sqrt{b},n)(k_2\sqrt{b},n)}{(\gamma,n)n!} \left(\frac{z}{b}\right)^n$$

The factor

$$\frac{(k_1\sqrt{b},n)(k_2\sqrt{b},n)}{b^n} = \frac{(k_1\sqrt{b},n)}{b^{n/2}} \frac{(k_2\sqrt{b},n)}{b^{n/2}} \to k_1^n k_2^n \text{ as } b \to \infty$$

by the same arguments as used in Section 7.1. We get

---

[5] Charles Hermite (1822–1901), French mathematician.

$$\lim_{b \to \infty} {}_2F_1(k_1\sqrt{b}, k_2\sqrt{b}; \gamma; z/b) = \sum_{n=0}^{\infty} \frac{(k_1 k_2 z)^n}{(\gamma, n) n!}$$

We introduce the notation

$$_0F_1(\gamma; z) = \sum_{n=0}^{\infty} \frac{1}{(\gamma, n)} \frac{z^n}{n!} \qquad (7.10)$$

This series converges everywhere in the complex plane, and the differential equation of $u(z) = {}_0F_1(\gamma; z)$ is

$$zu''(z) + \gamma u'(z) - u(z) = 0 \qquad (7.11)$$

which is obtained by taking the appropriate limit $b \to \infty$ in the scaled version ($\alpha = \beta = \sqrt{b}$, $z \to z/b$) of the hypergeometric equation (4.3) on page 50. This differential equation has one regular point at the origin, and one irregular point at infinity. Just like confluence of the first kind, confluence of the second kind has created an irregular singular point out of the regular singular point at $z = b \to \infty$ and the one located at infinity.

Just as for confluence of the first kind, the indicial equation, for the regular singular point at the origin, has roots 0 and $1 - \gamma$, respectively, and the second solution to the differential equation is

$$v(z) = z^{1-\gamma} f(z)$$

where, use Theorem 2.1 on page 4,

$$zf''(z) + (2 - \gamma)f'(z) - u(z) = 0$$

This differential equation has a solution, which is well behaved at the origin, viz.,

$$f(z) = {}_0F_1(2 - \gamma; z)$$

This observation implies that

$$v(z) = z^{1-\gamma} {}_0F_1(2 - \gamma; z), \quad \gamma \text{ not an integer}$$

is a second solution to the differential equation.

7.2.1 Integral representations

In parallel to Barnes' integral representation of the hypergeometric function in Theorem 5.4 on page 80, and the integral representation of the confluent hypergeometric function of the first kind in Theorem 7.2 on page 127, there is an integral representation of the confluent hypergeometric function of the second kind.

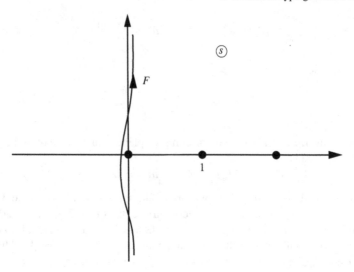

Fig. 7.2 The contour of integration in Theorem 7.3.

Theorem 7.3. *If* $\operatorname{Re}\gamma > 1$, *the confluent hypergeometric function of the second kind can be represented as a contour integral*

$$_0F_1(\gamma;z) = \frac{1}{2\pi i}\int_F \frac{\Gamma(\gamma)}{\Gamma(\gamma+s)}\Gamma(-s)(-z)^s\,ds, \quad \mathbb{R}\ni z<0$$

where F is a contour in the complex s-plane, starting at $-i\infty$ *and ending at* $i\infty$, *such that all poles to* $\Gamma(-s)$ *(i.e.,* $s\in\mathbb{N}$*) lie to the right of F, see Figure 7.2.*

Proof. The proof follows the proof of Theorem 7.2 closely. We start by rewriting the integrand, using (A.4), as

$$\frac{\Gamma(-s)(-z)^s}{\Gamma(\gamma+s)} = -\frac{(-z)^s}{\Gamma(\gamma+s)\Gamma(1+s)}\frac{\pi}{\sin\pi s}$$

The gamma functions are estimated using Corollary A.2 on page 172 with $\alpha = 1$ and $\alpha = \gamma$. The result is (C_1 is a constant independent of s)

$$\left|\frac{1}{\Gamma(\gamma+s)\Gamma(1+s)}\right| \leq C_1 e^{-\operatorname{Re}(2s+\gamma)\ln|s|+2\operatorname{Im}s\arg(s)+2\operatorname{Re}s}$$

Moreover, we have for $|s+n| > \varepsilon, \forall n\in\mathbb{Z}$, see Lemma A.1,

$$|\sin\pi s| \geq C_3 e^{\pi|\operatorname{Im}s|}$$

and also for real $z < 0$

$$|(-z)^s| = e^{\operatorname{Re}s\ln|z|}$$

The integrand at large values of s can now be estimated for real $z < 0$,

$$\left| \frac{\Gamma(-s)(-z)^s}{\Gamma(\gamma+s)} \right| \le Ce^{-\operatorname{Re}(\gamma+2s)\ln|s|+\operatorname{Re}s(2+\ln|z|)+2|\operatorname{Im}s|(|\arg(s)|-\pi/2)} \tag{7.12}$$

for $s \in \{s \in \mathbb{C} : |s+n| \ge \varepsilon, \forall n \in \mathbb{N}\}$. Notice that the factor $|\operatorname{Im}s|(|\arg(s)| - \pi/2)$ is bounded as $s \to \pm i\infty$. The convergence properties as $s \to \pm i\infty$ are determined by the factor

$$e^{-\operatorname{Re}\gamma\ln|s|}$$

and the integrand vanishes at least as $1/|s|$, as $s \to \pm i\infty$, since $\operatorname{Re}\gamma > 1$. The integrand is therefore convergent, and the first part of the proof is completed.

We next show that the integral in the theorem is identical to confluent hypergeometric function of the second kind in (7.10). As in the proof in Theorem 7.2, we close the contour by a semi-circle C in the right-hand s-plane. The poles at $s \in \mathbb{N}$ are avoided by parameterizing the contour C as

$$C: s = (N+1/2)e^{i\phi} = (N+1/2)(\cos\phi + i\sin\phi), \qquad \phi \in [-\pi/2, \pi/2]$$

and let $N \to \infty$ through the integer numbers. The integrand on C is bounded by, see (7.12),

$$\left| \frac{\Gamma(-s)(-z)^s}{\Gamma(\gamma+s)} \right|$$
$$\le Ce^{-(\operatorname{Re}\gamma+(2N+1)\cos\phi)\ln|N+1/2|+(N+1/2)\cos\phi(2+\ln|z|)+2(N+1/2)|\sin\phi|(|\phi|-\pi/2)}$$

and we observe that the convergence properties are determined by the factor

$$e^{-\operatorname{Re}\gamma\ln|N+1/2|}e^{-(N+1/2)\cos\phi\{2\ln|N+1/2|+2+\ln|z|\}}$$

which vanishes at least as $N^{-\operatorname{Re}\gamma}$ as $N \to \infty$, showing that the contour C does not contribute in the limit as $N \to \infty$. The residue theorem implies

$$\int_F \frac{\Gamma(\gamma)}{\Gamma(\gamma+s)}\Gamma(-s)(-z)^s\,ds = \int_{F-C} \frac{\Gamma(\gamma)}{\Gamma(\gamma+s)}\Gamma(-s)(-z)^s\,ds$$

$$= -2\pi i \sum_{n=0}^{\infty} \operatorname*{Res}_{s=n} \frac{\Gamma(\gamma)}{\Gamma(\gamma+s)}\Gamma(-s)(-z)^s$$

The residues of $\Gamma(-s)$, see (5.15) on page 84, then imply

$$\int_F \frac{\Gamma(\gamma)}{\Gamma(\gamma+s)}\Gamma(-s)(-z)^s\,ds = 2\pi i \sum_{n=0}^{\infty} \frac{\Gamma(\gamma)}{\Gamma(\gamma+n)}\frac{(-1)^n}{n!}(-z)^n$$

$$= 2\pi i \sum_{n=0}^{\infty} \frac{1}{(\gamma,n)}\frac{z^n}{n!} = 2\pi i {}_0F_1(\gamma;z)$$

and the theorem is proved. ☐

7.2.2 Bessel functions — revisited

The Bessel functions, $J_v(z)$, were introduced in Section 7.1.1 and were found to be related to the confluent hypergeometric function $_1F_1(\gamma;z)$, see (7.5),

$$J_v(z) = \frac{z^v e^{-iz} {}_1F_1(v+1/2; 2v+1; 2iz)}{2^v \Gamma(v+1)}$$

In this section, we show that $J_v(z)$ also can be expressed in the confluent hypergeometric function of the second kind, $_0F_1(\gamma;z)$.

The Bessel function $J_v(z)$ satisfies the Bessel differential equation, see (7.3),

$$z^2 u''(z) + z u'(z) + (z^2 - v^2) u(z) = 0$$

and the Bessel function $J_v(z)$ behaves near the origin as

$$J_v(z) \sim \frac{z^v}{2^v \Gamma(v+1)}$$

The derivation follows the analysis in Section 7.1.1 closely, and we start with the confluent hypergeometric equation, (7.11)

$$z u''(z) + \gamma u'(z) - u(z) = 0$$

As in Section 7.1.1, we use Theorem 2.1 on page 4 but now $g(t) = -t^2/4$. The confluent hypergeometric equation above then is transformed into

$$t w''(t) + (2\gamma - 1) w'(t) + t w(t) = 0$$

Another transformation with Theorem 2.1 with $w(t) = t^{-v} u(t)$ gives

$$t^2 u''(t) + (2(\gamma - v - 1) + 1) t u'(t) + (t^2 - 2v(\gamma - v - 1) - v^2) u(t) = 0$$

From these transformations, we see that by choosing $\gamma = v + 1$, the equation is identical to the Bessel differential equation, leading to the identification of the Bessel function, $J_v(z)$, in terms of the confluent hypergeometric function, $_0F_1(\gamma;z)$

$$J_v(z) = \frac{z^v {}_0F_1(v+1; -z^2/4)}{2^v \Gamma(v+1)} \tag{7.13}$$

with power series expansion

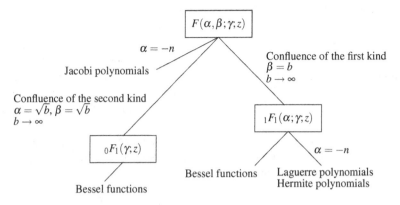

Fig. 7.3 An overview of the hierarchy of the solutions to the hypergeometric differential equation and its two confluent tracks.

$$J_v(z) = \left(\frac{z}{2}\right)^v \sum_{n=0}^{\infty} \frac{(-1)^n}{\Gamma(n+v+1)\,n!} \left(\frac{z}{2}\right)^{2n}$$

We conclude the section with an example.

Example 7.2. From (7.13) and Theorem 7.3, we get ($x > 0$ and $\operatorname{Re} v > 0$)

$$J_v(x) = \frac{x^v {}_0F_1(v+1; -x^2/4)}{2^v \Gamma(v+1)}$$

$$= \frac{x^v}{2^v \Gamma(v+1)} \frac{1}{2\pi i} \int_F \frac{\Gamma(v+1)}{\Gamma(v+1+s)} \Gamma(-s)(x^2/4)^s \, ds$$

where the path of integration goes along the imaginary axis, but to the left of the singular point at $s = 0$. We get the integral representation

$$J_v(x) = \frac{1}{2\pi i} \int_{-i\infty}^{i\infty} \frac{\Gamma(-s)(x/2)^{v+2s}}{\Gamma(v+1+s)} \, ds, \quad x > 0, \quad \operatorname{Re} v > 0$$

■

7.3 Solutions with three singular points — a summary

After having explored the solutions of the hypergeometric function ${}_2F_1(\alpha, \beta; \gamma; z)$ in Chapter 5, and its confluent versions, ${}_1F_1(\alpha; \gamma; z)$ and ${}_0F_1(\gamma; z)$, in this chapter, it is appropriate to illustrate the overall structure of the solutions. To this end, we collect the different solutions in Figure 7.3.

7.4 Generalized hypergeometric series

For completeness, we end this chapter with the definition of the generalized hyper-geometric series of $_mF_n(\alpha_1,\ldots,\alpha_m;\gamma_1,\ldots,\gamma_n;z)$. We do not give any detailed description of the properties of this function in this section, but many of its properties can be derived using techniques similar to the ones used above for the hypergeometric function and the confluent hypergeometric functions of the first and second kind.

The generalized hypergeometric series for general indices m and n has terms with m factors in the nominator and n factors in the denominator of the series representation of the solution. Its definition is

$$_mF_n(\alpha_1,\ldots,\alpha_m;\gamma_1,\ldots,\gamma_n;z) = \sum_{k=0}^{\infty} \frac{(\alpha_1,k)\cdot\ldots\cdot(\alpha_m,k)}{(\gamma_1,k)\cdot\ldots\cdot(\gamma_n,k)} \frac{z^k}{k!} \tag{7.14}$$

The convergence properties of this series depend on how the m and n indices are related to each other. By the ratio test we find

$$\begin{cases} m \leq n \Longrightarrow \text{convergence for } |z| < \infty \\ m = n+1 \Longrightarrow \text{convergence for } |z| < 1 \\ m > n+1 \Longrightarrow \text{divergence everywhere} \end{cases}$$

Problems

7.1. Show that

$$_1F_1(\alpha;\gamma;z) = e^z {_1F_1}(\gamma-\alpha;\gamma;-z)$$

7.2. Prove

$$\frac{d}{dz} {_1F_1}(\alpha;\gamma;z) = \frac{\alpha}{\gamma} {_1F_1}(\alpha+1;\gamma+1;z)$$

and more generally

$$\frac{d^n}{dz^n} {_1F_1}(\alpha;\gamma;z) = \frac{(\alpha,n)}{(\gamma,n)} {_1F_1}(\alpha+n;\gamma+n;z)$$

7.3. Determine the Wronskian of the two linearly independent solutions $_1F_1(\alpha;\gamma;z)$ and $v(z)$ in (7.1) and (7.2), respectively.

7.4. The Whittaker[6] equation reads

$$z^2 u''(z) + \left(\frac{1}{4} - \mu^2 + \kappa z - \frac{z^2}{4}\right) u(z) = 0$$

---

[6] Edmund Taylor Whittaker (1873–1956), English mathematician.

Find its solutions in terms of the Kummer function $_1F_1(\alpha;\gamma;z)$.

7.5. Weber's[7] equation reads

$$u''(z) + \left(v + \frac{1}{2} - \frac{z^2}{4}\right)u(z) = 0$$

Its solutions are often called parabolic cylinder functions. Find its solutions in terms of the Kummer function $_1F_1(\alpha;\gamma;z)$.

7.6. Show that for any $\sigma < 0$

$$e^z = \frac{1}{2\pi i}\int_{\sigma-i\infty}^{\sigma+i\infty}\Gamma(-s)(-z)^s\,ds, \qquad \mathrm{Re}\,z < 0$$

7.7. The error function $\mathrm{erf}(z)$ is defined as

$$\mathrm{erf}(z) = \frac{2}{\sqrt{\pi}}\int_0^z e^{-t^2}\,dt$$

Show that

$$\mathrm{erf}(z) = \frac{2z}{\sqrt{\pi}}\,_1F_1(1/2;3/2;-z^2) = \frac{2z}{\sqrt{\pi}}e^{-z^2}\,_1F_1(1;3/2;z^2)$$

---

[7] Heinrich Friedrich Weber (1843–1912), German physicist.

Chapter 8
Heun's differential equation

The solutions to the hypergeometric differential equation — two regular singular points in the finite complex plane (at $z = 0$ and $z = 1$) and a regular singular point at infinity — were analyzed in detail in Chapter 5. In this chapter, the solutions of the differential equation with four regular singular points are investigated. We restrict ourselves to the situation where there are three regular singular points in the finite complex plane, and a regular singular point at infinity.

8.1 Basic properties

The differential equation with three regular singular points in the finite complex plane and one regular singular point at infinity was introduced in Section 4.4.1. The three regular singular points in the finite complex plane are located at $z = 0$, $z = 1$, and $z = a$, respectively. We restrict ourselves to the case when $|a| > 1$ (no restriction if $|a| \neq 1$, since the role of the singular points $z = 1$ and $z = a$ can be changed), see Figure 8.1. The differential equation that corresponds to these conditions is Heun's differential equation, which reads, see (4.10) on page 57,

$$u''(z) + \left[\frac{\gamma}{z} + \frac{\delta}{z-1} + \frac{\varepsilon}{z-a}\right] u'(z) + \frac{\alpha\beta(z-h)}{z(z-1)(z-a)} u(z) = 0 \qquad (8.1)$$

The extra condition that the coefficients have to satisfy is

$$\begin{cases} \alpha + \beta + 1 = \gamma + \delta + \varepsilon \\ h \text{ accessory parameter} \end{cases} \qquad (8.2)$$

The roots of the indicial equations at $z = 0$ are 0 and $1 - \gamma$. Similarly, the roots at $z = 1$ are 0 and $1 - \delta$, at $z = a$ they are 0 and $1 - \varepsilon$, and, finally, at $z = \infty$ they are α and β. The roots of the indicial equations at the four different regular singular points are summarized in Table 8.1.

G. Kristensson, *Second Order Differential Equations: Special Functions and Their Classification*, DOI 10.1007/978-1-4419-7020-6_8,
© Springer Science+Business Media, LLC 2010

Table 8.1 The roots of the indicial equation of Heun's differential equation. The parameters α, β γ, δ, and ε satisfy (8.2).

Point	Roots λ
$z = 0$	$0, 1 - \gamma$
$z = 1$	$0, 1 - \delta$
$z = a$	$0, 1 - \varepsilon$
$z = \infty$	α, β

Generalizing the notation by Riemann introduced in Section 4.3.2, see also (4.11), we write the set of solutions to Heun's differential equation as

$$
u(z) \in P \left\{ \begin{matrix} 0 & 1 & a & \infty \\ 0 & 0 & 0 & \alpha \ z \\ 1-\gamma & 1-\delta & 1-\varepsilon & \beta \end{matrix} \right\}
$$

In Section 5.4, we concluded that the solutions of the hypergeometric differential equation show symmetry properties with respect to transformations of the independent variable z, see Theorem 5.2. A similar, but more complex result, holds for the solutions to Heun's differential equation. In total, there are 192 different solutions to Heun's differential equation, corresponding to the 24 permutations of the regular singular points $\{0, 1, a, \infty\}$ and the 8 combinations of the exponential factors corresponding to the roots of the indicial equation, $\{\{0, 1 - \gamma\}, \{0, 1 - \delta\}, \{0, 1 - \varepsilon\}\}$, see also Problem 4.5. The result relies on the generalization of Theorem 4.1 on page 52 to four singular points, i.e., with the notation of Theorem 4.1

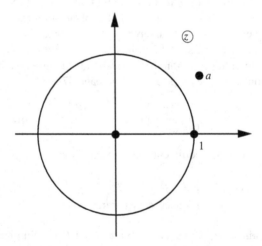

Fig. 8.1 The three regular singular points at $z = 0, 1, a$ in Heun's differential equation. The fourth regular singular point is located at infinity.

$$P\left\{\begin{matrix} a_1 \ a_2 \ a_3 \ a_4 \\ \alpha_1 \ \alpha_2 \ \alpha_3 \ \alpha_4 \ z \\ \beta_1 \ \beta_2 \ \beta_3 \ \beta_4 \end{matrix}\right\} = P\left\{\begin{matrix} t(a_1) \ t(a_2) \ t(a_3) \ t(a_4) \\ \alpha_1 \quad \alpha_2 \quad \alpha_3 \quad \alpha_4 \ t(z) \\ \beta_1 \quad \beta_2 \quad \beta_3 \quad \beta_4 \end{matrix}\right\}$$

This result is a consequence of Problem 3.5. We refrain from giving the details of this result here, but refer to Ronveaux [21] and Maier [16] for a comprehensive treatment of the topic.

Comment 8.1. Heun's differential equation reduces to the hypergeometric differential equation if we let $a = 1$ (confluence) and $h = 1$. To see this, rewrite Heun's differential equation in the form

$$z(z-1)^2 u''(z) + z(z-1)^2 \left[\frac{\gamma}{z} + \frac{\delta+\varepsilon}{z-1}\right] u'(z) + \alpha\beta(z-1)u(z) = 0$$

or

$$z(z-1)u''(z) + z(z-1)\left[\frac{\gamma}{z} + \frac{\alpha+\beta-\gamma+1}{z-1}\right] u'(z) + \alpha\beta u(z) = 0$$

which is the hypergeometric differential equation

$$z(z-1)u''(z) + [(\alpha+\beta+1)z - \gamma]u'(z) + \alpha\beta u(z) = 0$$

∎

8.2 Power series solution

Just as in the analysis of the solutions to the hypergeometric differential equation in Chapter 5, we develop power series solutions to Heun's differential equation at $z = 0$, i.e., we make the ansatz

$$u(z) = \sum_{n=0}^{\infty} a_n z^n, \qquad v(z) = z^{1-\gamma} \sum_{n=0}^{\infty} a'_n z^n \qquad (8.3)$$

Insert the first ansatz in (8.3) into Heun's differential equation and identify the coefficient in front of the nth power ($n \in \mathbb{Z}_+$). The result is

$$(n-1)(n-2)a_{n-1} - (1+a)n(n-1)a_n + a(n+1)na_{n+1} + (\gamma+\delta+\varepsilon)(n-1)a_{n-1}$$
$$- (\gamma+\gamma a+\delta a+\varepsilon)na_n + \gamma a(n+1)a_{n+1} + \alpha\beta a_{n-1} - \alpha\beta h a_n = 0$$

or with the use of the constraint condition in (8.2)

$$(n-1+\alpha)(n-1+\beta)a_{n-1} - [(1+a)n(n-1+\gamma) + (\delta a+\varepsilon)n + \alpha\beta h]a_n$$
$$+ a(n+1)(n+\gamma)a_{n+1} = 0$$

The coefficient corresponding to $n = 0$ is

$$a\gamma a_1 - \alpha\beta h a_0 = 0$$

We write the relation between the coefficients in the following way:

$$a_{n+1} = A_n a_n + B_n a_{n-1}, \quad n \in \mathbb{Z}_+ \tag{8.4}$$

where

$$
\begin{cases}
A_n = \dfrac{(1+a)n(n-1+\gamma) + (\delta a + \varepsilon)n + \alpha\beta h}{a(n+1)(n+\gamma)} \\[4mm]
B_n = -\dfrac{(n-1+\alpha)(n-1+\beta)}{a(n+1)(n+\gamma)}
\end{cases}
\quad n \in \mathbb{N} \tag{8.5}
$$

The sequence is initialized by

$$a_1 = \frac{\alpha\beta h}{a\gamma} a_0 = A_0 a_0$$

The values of $\gamma = 0, -1, -2, \ldots$ have to be excluded from the analysis, otherwise the coefficients a_n become undetermined at some n value. The coefficients A_n and B_n behave asymptotically as $n \to \infty$ as

$$
\begin{cases}
A_n = 1 + \dfrac{1}{a} + \dfrac{-2a - 2 + \delta a + \varepsilon}{an} + O(1/n^2) \\[4mm]
B_n = -\dfrac{1}{a} - \dfrac{\alpha + \beta - \gamma - 3}{an} + O(1/n^2) = -\dfrac{1}{a} - \dfrac{\delta + \varepsilon - 4}{an} + O(1/n^2)
\end{cases} \tag{8.6}
$$

The convergence of the power series solution in (8.3) is determined by studying the behavior of the quotient a_{n+1}/a_n for large n. The limit

$$t = \lim_{n \to \infty} \frac{a_{n+1}}{a_n}$$

exists by the results in Appendix B.2, see Theorem B.2 on page 183. With $A = 1 + 1/a$ and $B = -1/a$, the roots to the characteristic equation (B.3) satisfy

$$t^2 = t\frac{1+a}{a} - \frac{1}{a}$$

with solutions

$$
\begin{cases}
t_1 = 1/a \\
t_2 = 1
\end{cases}
$$

Here $|t_2| > |t_1|$ by the assumption made on a. The radius of convergence of the power series, (8.3), then is $r = 1/|t|$. As shown in Theorem B.2, the proper root is t_2, thus the power series converges absolutely inside the unit circle, $r = 1$, in the complex z-plane, see Figure 8.2, and diverges outside this circle. In the exceptional case, see (B.4) in Appendix B.2, the quotient a_{n+1}/a_n converges to t_1, thus leading to a larger domain of convergence, i.e., $r = |a|$. Provided this is the case, power series converges absolutely for $|z| < |a|$ and divergences for $|z| > |a|$.

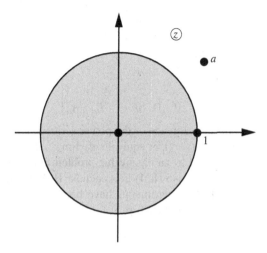

Fig. 8.2 The domain of convergence of the power series expansion of the solution to Heun's differential equation that is analytic at the origin.

The convergence on the unit circle, $|z| = 1$, is investigated by the use of Raabe's test, see footnote 9 on page 64. In Problem 8.1 it is proved that the power series converges absolutely provided $\operatorname{Re}\delta < 1$. In the exceptional case, the power series converges on the circle $|z| = |a|$ if $\operatorname{Re}\varepsilon < 1$, see Problem 8.1.

We have seen that the power series (8.3) converges inside the unit circle. We adopt the normalization $a_0 = 1$, and call the solution obtained by the power series a local solution, and use the notion

$$u(z) = Hl(a, h; \alpha, \beta, \gamma, \delta, \varepsilon; z)$$

We summarize these conclusions made so far in this section in a theorem.

Theorem 8.1. *There is a solution $u(z)$ to Heun's differential equation, (8.1), analytic at the origin and $u(0) = 1$, with a power series expansion that converges absolutely inside the unit circle. Moreover, the power series converges absolutely on the unit circle provided $\operatorname{Re}\delta < 1$. This solution is denoted $u(z) = Hl(a, h; \alpha, \beta, \gamma, \delta, \varepsilon; z)$.*

In the exceptional case, see (B.4), the power series has a larger domain of convergence, i.e., $r = |a|$. The power series converges on the circle $|z| = |a|$ provided $\operatorname{Re}\varepsilon < 1$.

We conclude this section with a brief discussion of the special conditions that have to be met in order to fulfill the exceptional case. Lemma B.1 in Appendix B shows that the coefficients a_n can be determined as a determinant D_n, i.e., $a_n = D_n a_0$, where

$$D_{n+1} = \begin{vmatrix} A_0 & -1 & 0 & 0 & 0 & \cdots & \cdots & \cdots \\ B_1 & A_1 & -1 & 0 & 0 & \cdots & \cdots & \cdots \\ 0 & B_2 & A_2 & -1 & 0 & \cdots & \cdots & \cdots \\ \vdots & \vdots & \vdots & \vdots & \vdots & \cdots & & \vdots \\ \cdots & \cdots & \cdots & \cdots & \cdots & \cdots & A_{n-1} & -1 \\ 0 & 0 & 0 & 0 & 0 & \cdots & B_n & A_n \end{vmatrix} \qquad n \in \mathbb{N} \qquad (8.7)$$

A necessary condition for convergence of the power series in a larger domain than the unit circle[1] is that $\lim_{n\to\infty} a_n = 0$, or equivalently $\lim_{n\to\infty} D_n = 0$. This condition for convergence can be viewed as an eigenvalue problem implying a condition on the accessory parameter h, see also (B.4) in Appendix B. We number the allowed values as $h_1, h_2, h_3, \ldots$. Infinite determinants have been studied by Hill,[2] and we refer to the literature for more details [31, p. 36].

8.3 Polynomial solution

Polynomial solutions play an important role, and in this section the conditions for a polynomial solution of Heun's differential equation are investigated.

The power series in (8.3) is terminated, resulting in a polynomial solution of degree n, if $D_{n+1} = 0$ in (8.7) and, at the same time, $a_n \neq 0$. To terminate the sequence in (8.4) after n terms, we also need $B_{n+1} = 0$, which implies

$$(n+\alpha)(n+\beta) = 0 \implies \alpha = -n \text{ or } \beta = -n$$

Example 8.1. The simplest example is $n = 1$, which implies that $\alpha = -1$ or $\beta = -1$. In both cases $\alpha\beta = -(\gamma + \delta + \varepsilon)$, due to the extra condition $\alpha + \beta + 1 = \gamma + \delta + \varepsilon$.

$$D_2 = 0 \implies D_2 = \begin{vmatrix} A_0 & -1 \\ B_1 & A_1 \end{vmatrix} = A_0 A_1 + B_1 = 0$$

We get

$$\frac{\alpha\beta h}{a\gamma} \frac{(1+a)\gamma + \delta a + \varepsilon + \alpha\beta h}{2a(1+\gamma)} - \frac{\alpha\beta}{2a(1+\gamma)} = 0$$

The solution has to satisfy

$$0 = h(1+a)\gamma + h\delta a + h\varepsilon - h^2(\gamma + \delta + \varepsilon) - a\gamma$$
$$= (a-h)(h-1)\gamma + h(a-h)\delta + h(1-h)\varepsilon$$

or

$$\frac{\gamma}{h} + \frac{\delta}{h-1} + \frac{\varepsilon}{h-a} = 0$$

---

[1] The necessary condition for convergence of the power series is that $|a_n||z|^n \to 0$ as $n \to \infty$, but this implies $|a_n| \to 0$ as $n \to \infty$ if $|z| > 1$.

[2] George William Hill (1838–1914), U.S. astronomer and mathematician.

Given γ, δ, and ε, this is a quadratic equation in h to solve. This gives two polynomials $1 + a_1 z$, where

$$a_1 = \frac{\alpha\beta h}{a\gamma} = -h\frac{\gamma + \delta + \varepsilon}{a\gamma}$$

∎

8.4 Solution in hypergeometric polynomials

A power series solution of Heun's differential equations was analyzed in Section 8.2. It was proved that this solution was absolutely convergent inside the unit circle, $|z| < 1$. A rearrangement of the terms in the power series suggests another type of solution expressed as an infinite series of polynomials. The hypergeometric polynomials (Jacobi polynomials) in Section 5.8, $P_n^{(\alpha,\beta)}(x)$, are polynomials of degree n, and a suitable candidate for these polynomials. This type of solution was first investigated by N. Svartholm [25]. In this section, we investigate under what conditions this series of the hypergeometric polynomials (Jacobi polynomials) is a solution of Heun's differential equation.

To this end, make the following ansatz in hypergeometric polynomials (Jacobi polynomials)

$$u(z) = \sum_{n=0}^{\infty} c_n y_n(z) \tag{8.8}$$

where we have introduced a short-hand notation for the Jacobi polynomials, see also (5.18) on page 94

$$y_n(z) = F(-n, n + \gamma + \delta - 1; \gamma; z) = \frac{n!\,\Gamma(\gamma)}{\Gamma(n + \gamma)} P_n^{(\gamma - 1, \delta - 1)}(1 - 2z)$$

and with $\omega = \gamma + \delta - 1$, we have

$$y_n(z) = F(-n, n + \omega; \gamma; z)$$

The first two functions are

$$\begin{cases} y_0(z) = 1 \\ y_1(z) = 1 - \dfrac{1 + \omega}{\gamma} z \end{cases}$$

8.4.1 Asymptotic properties of the polynomials $y_n(z)$

The asymptotic behavior of the polynomials $y_n(z)$ for large indices n is needed to determine the domain of convergence of the series in (8.8). This section is devoted to an investigation of these properties.

The polynomials $y_n(z)$ satisfy, see Section 5.1,

$$y_n''(z) + \left[\frac{\gamma}{z} + \frac{\delta}{z-1}\right] y_n'(z) - \frac{n(n+\omega)}{z(z-1)} y_n(z) = 0 \tag{8.9}$$

and the recursion relations, see Problem 8.2, $n \in \mathbb{Z}_+$

$$z y_n(z) = P_n y_{n+1}(z) + Q_n y_n(z) + R_n y_{n-1}(z) \tag{8.10}$$

$$z(z-1)\frac{d}{dz} y_n(z) = P_n' y_{n+1}(z) + Q_n' y_n(z) + R_n' y_{n-1}(z) \tag{8.11}$$

where $(n \in \mathbb{Z}_+)$

$$
\begin{cases}
P_n = -\dfrac{(n+\omega)(n+\gamma)}{(2n+\omega)(2n+\omega+1)} \\[2mm]
Q_n = \dfrac{(\omega-1)(\gamma-\delta)}{2(2n+\omega+1)(2n+\omega-1)} + \dfrac{1}{2} \\[2mm]
R_n = -\dfrac{n(n+\delta-1)}{(2n+\omega)(2n+\omega-1)}
\end{cases}
\qquad
\begin{cases}
P_n' = -\dfrac{n(n+\omega)(n+\gamma)}{(2n+\omega)(2n+\omega+1)} \\[2mm]
Q_n' = \dfrac{n(n+\omega)(\gamma-\delta)}{(2n+\omega-1)(2n+\omega+1)} \\[2mm]
R_n' = \dfrac{n(n+\omega)(n+\delta-1)}{(2n+\omega)(2n+\omega-1)}
\end{cases}
$$

with initializing values $(n = 0)$

$$
\begin{cases}
P_0 = -\dfrac{\gamma}{\omega+1} \\[2mm]
Q_0 = \dfrac{\gamma}{\omega+1} \\[2mm]
R_0 = 0
\end{cases}
\qquad
\begin{cases}
P_0' = 0 \\[1mm]
Q_0' = 0 \\[1mm]
R_0' = 0
\end{cases}
$$

The dominant contributions for large index values as $n \to \infty$ are

$$
\begin{cases}
P_n = -\dfrac{1}{4} + \dfrac{1-2\gamma}{8n} + O(1/n^2) \\[2mm]
Q_n = \dfrac{1}{2} + O(1/n^2) \\[2mm]
R_n = -\dfrac{1}{4} - \dfrac{1-2\gamma}{8n} + O(1/n^2)
\end{cases}
\qquad
\begin{cases}
P_n' = -\dfrac{n}{4} + \dfrac{1-2\gamma}{8} + O(1/n) \\[2mm]
Q_n' = \dfrac{\gamma-\delta}{4} + O(1/n^2) \\[2mm]
R_n' = \dfrac{n}{4} - \dfrac{1-2\delta}{8} + O(1/n)
\end{cases} \tag{8.12}
$$

The main result of the asymptotic behavior is contained in the following lemma:

Lemma 8.1. *The limit*

$$\lim_{n\to\infty} \frac{y_{n+1}(z)}{y_n(z)} = s(z) = \frac{\left(1-z^{-1}\right)^{1/2}+1}{\left(1-z^{-1}\right)^{1/2}-1}$$

exists, where the branch of the square root satisfies $\operatorname{Re}\left(1-z^{-1}\right)^{1/2} > 0$ (we assume $z \notin [0,1]$). The function $s(z)$ also satisfies $|s(z)| > 1$. Moreover,

$$\frac{y_{n+1}(z)}{y_n(z)} = s(z) + \frac{\sigma(z)}{n} + O(1/n^2), \quad as\ n\to\infty$$

where $\sigma(z)$ is

$$\sigma(z) = s(z)\frac{1-2\gamma}{2}$$

Proof. We make use of the recursion relation for the functions $y_n(z)$ in (8.10).

$$zy_n(z) = P_n y_{n+1}(z) + Q_n y_n(z) + R_n y_{n-1}(z)$$

The coefficients, P_n, Q_n, R_n, have well-defined limits as $n\to\infty$. From (8.12) we have

$$\begin{cases} P_n = -\dfrac{1}{4} + \dfrac{1-2\gamma}{8n} + O(1/n^2) \\[2mm] Q_n = \dfrac{1}{2} + O(1/n^2) \\[2mm] R_n = -\dfrac{1}{4} - \dfrac{1-2\gamma}{8n} + O(1/n^2) \end{cases} \qquad as\ n\to\infty$$

To show that $y_{n+1}(z)/y_n(z)$ has a limit as $n\to\infty$, we can employ Theorem B.2 on page 183 directly. However, in this proof, we proceed in a somewhat different way. To this end, write

$$a_n(z) = s(z)^{-n} y_n(z)$$

where we choose $s(z)$ such that the new recursion relation

$$a_{n+1} = A_n a_n + B_n a_{n-1}$$

where

$$\begin{cases} A_n = \dfrac{z-Q_n}{sP_n} \\[3mm] B_n = -\dfrac{R_n}{s^2 P_n} \end{cases}$$

satisfies

$$\begin{cases} A_n = 1 + \alpha + O(1/n) \\ B_n = -\alpha + O(1/n) \end{cases}$$

with $|\alpha| < 1$. The asymptotic behavior of the coefficients, which is

$$\begin{cases} A_n = s^{-1}(2-4z) + O(1/n) \\ B_n = -s^{-2} + O(1/n) \end{cases}$$

leads us to choose s satisfying (remember we require $A_n + B_n \to 1$ as $n \to \infty$)

$$s^2 - 2s(1-2z) + 1 = 0 \tag{8.13}$$

with solutions

$$s_{1,2}(z) = 1 - 2z \pm 2\left(z^2 - z\right)^{1/2} = -\left(z^{1/2} \mp (z-1)^{1/2}\right)^2$$

which we write as

$$s_1(z) = \frac{Z-1}{Z+1}, \qquad s_2(z) = \frac{1}{s_1(z)} = \frac{Z+1}{Z-1}$$

where

$$Z = \left(1 - z^{-1}\right)^{1/2}$$

The first root satisfies

$$|s_1(z)| = \left|\frac{Z-1}{Z+1}\right| < 1$$

since $Z = \left(1 - z^{-1}\right)^{1/2}$ is located in the right-hand side of the complex Z-plane ($\operatorname{Re} Z > 0$), whenever $z \notin [0,1]$. Therefore, $|s_2(z)| > 1$, and in order to satisfy the condition $|\alpha| = |s(z)|^{-2} < 1$, we have to pick the root $s(z) = s_2(z)$.

We can now apply Corollary B.1 on page 193, which shows that the quotient a_{n+1}/a_n converges to 1, and, hence, the limit[3]

$$\lim_{n \to \infty} \frac{y_{n+1}(z)}{y_n(z)} = s_2(z) = s(z)$$

The first part of the lemma is then proved.

To prove the conditions on $\sigma(z)$, we once again make use of the recursion relation in (8.10) and divide by $y_n(z)$. The asymptotic behavior of the coefficients then implies

$$z = \left(-\frac{1}{4} + \frac{1-2\gamma}{8n} + O(1/n^2)\right)\left(s(z) + \frac{\sigma(z)}{n} + O(1/n^2)\right) + \frac{1}{2} + O(1/n^2)$$
$$+ \left(-\frac{1}{4} - \frac{1-2\gamma}{8n} + O(1/n^2)\right) \frac{1}{s(z) + \frac{\sigma(z)}{n-1} + O(1/n^2)}$$

---

[3] The Hungarian-born British mathematician Arthur Erdélyi (1908–1977) showed that the exceptional case

$$\lim_{n \to \infty} \frac{y_{n+1}(z)}{y_n(z)} = s_1(z) = \frac{1}{s(z)}$$

does not occur [7].

Use (8.13), and keep terms up to order $1/n$. We get

$$0 = -\frac{\sigma(z)}{4n} + s(z)\frac{1-2\gamma}{8n} + \frac{1}{s(z)}\left(\frac{\sigma(z)}{4s(z)(n-1)} - \frac{1-2\gamma}{8n}\right) + O(1/n^2)$$

A multiplication by n and taking the limit $n \to \infty$ give

$$2\sigma(z) = \left(s(z) - \frac{1}{s(z)}\right)(1-2\gamma) + 2\frac{\sigma(z)}{(s(z))^2}$$

which we solve for $\sigma(z)$. Finally, we get

$$\sigma(z) = s(z)\frac{1-2\gamma}{2}$$

which proves the lemma.   □

8.4.2 Asymptotic properties of the coefficients c_n

In a similar manner, we investigate the asymptotic behavior of the expansion coefficients c_n in the series (8.8), which satisfy Heun's differential equation. We collect the results in two lemmas.

Lemma 8.2. *If the series* (8.8) *is a solution of Heun's differential equation, the coefficients in the series of hypergeometric polynomials, c_n, satisfy*

$$c_{n+1} = E_n c_n + F_n c_{n-1}$$

where

$$\begin{cases} E_n = -\dfrac{\varepsilon Q_n' + \alpha\beta Q_n + n(n+\omega)Q_n - (\alpha\beta h + n(n+\omega)a)}{\varepsilon R_{n+1}' + (\alpha\beta + (n+1)(n+1+\omega))R_{n+1}} \\[4mm] F_n = -\dfrac{\varepsilon P_{n-1}' + \alpha\beta P_{n-1} + (n-1)(n-1+\omega)P_{n-1}}{\varepsilon R_{n+1}' + (\alpha\beta + (n+1)(n+1+\omega))R_{n+1}} \end{cases}$$

The sequence is initialized by

$$c_1 = E_0 c_0 = -\frac{\alpha\beta(2+\omega)(\gamma - h(\omega+1))}{\delta((\varepsilon-1)(\omega+1) - \alpha\beta)}c_0$$

Moreover, the coefficients E_n and F_n behave asymptotically as

$$\begin{cases} E_n = 2 - 4a + \dfrac{(1-2a)(2\gamma+2\varepsilon-5)}{n} + O(1/n^2) \\[4mm] F_n = -1 - \dfrac{2\gamma+2\varepsilon-5}{n} + O(1/n^2) \end{cases} \qquad \text{as } n \to \infty \qquad (8.14)$$

Proof. Insert the series expansion (8.8) in Heun's differential equation

$$u''(z) + \left[\frac{\gamma}{z} + \frac{\delta}{z-1} + \frac{\varepsilon}{z-a}\right] u'(z) + \frac{\alpha\beta(z-h)}{z(z-1)(z-a)} u(z) = 0$$

and use the differential equation for the function $y_k(z)$, see (8.9). We get

$$\varepsilon z(z-1) \sum_{k=0}^{\infty} c_k y_k'(z) + \sum_{k=0}^{\infty} c_k \left(\alpha\beta(z-h) + k(k+\omega)(z-a)\right) y_k(z) = 0$$

or using (8.10) and (8.11)

$$\varepsilon \sum_{k=0}^{\infty} c_k \left(P_k' y_{k+1}(z) + Q_k' y_k(z) + R_k' y_{k-1}(z)\right)$$

$$+ \sum_{k=0}^{\infty} c_k \left(\alpha\beta + k(k+\omega)\right) \left(P_k y_{k+1}(z) + Q_k y_k(z) + R_k y_{k-1}(z)\right)$$

$$- \sum_{k=0}^{\infty} c_k \left(\alpha\beta h + k(k+\omega)a\right) y_k(z) = 0$$

Since the hypergeometric polynomials form an orthogonal set, see Lemma 5.8, the coefficient in front of $y_n(z)$ must be zero, leading to ($n \in \mathbb{Z}_+$)

$$\varepsilon \left(c_{n-1} P_{n-1}' + c_n Q_n' + c_{n+1} R_{n+1}'\right) + \alpha\beta \left(c_{n-1} P_{n-1} + c_n Q_n + c_{n+1} R_{n+1}\right)$$
$$+ c_{n-1}(n-1)(n-1+\omega)P_{n-1} + c_n n(n+\omega)Q_n + c_{n+1}(n+1)(n+1+\omega)R_{n+1}$$
$$- (\alpha\beta h + n(n+\omega)a) c_n = 0$$

The coefficient in front of $y_0(z)$ is used as an initialization of recursion relation (use $P_0' = Q_0' = R_0' = 0$)

$$\varepsilon c_1 R_1' + \alpha\beta (c_0 Q_0 + c_1 R_1) + c_1(1+\omega)R_1 - \alpha\beta h c_0 = 0$$

or explicitly

$$\delta\left(\frac{\varepsilon-1}{2+\omega} - \frac{\alpha\beta}{(2+\omega)(\omega+1)}\right) c_1 + \alpha\beta\left(\frac{\gamma}{\omega+1} - h\right) c_0 = 0$$

or

$$c_1 = -\frac{\alpha\beta(2+\omega)(\gamma - h(\omega+1))}{\delta((\varepsilon-1)(\omega+1) - \alpha\beta)} c_0$$

We write the recursion relation as

$$c_{n+1} = E_n c_n + F_n c_{n-1}$$

where

$$\begin{cases} E_n = -\dfrac{\varepsilon Q'_n + \alpha\beta Q_n + n(n+\omega)Q_n - (\alpha\beta h + n(n+\omega)a)}{\varepsilon R'_{n+1} + (\alpha\beta + (n+1)(n+1+\omega))R_{n+1}} \\[3mm] F_n = -\dfrac{\varepsilon P'_{n-1} + \alpha\beta P_{n-1} + (n-1)(n-1+\omega)P_{n-1}}{\varepsilon R'_{n+1} + (\alpha\beta + (n+1)(n+1+\omega))R_{n+1}} \end{cases}$$

The asymptotic behavior in (8.14) is obtained by the use of (8.12). The denominator behaves as

$$\varepsilon R'_{n+1} + (\alpha\beta + (n+1)(n+1+\omega))R_{n+1}$$
$$= -\frac{n^2}{4} + \frac{n}{4}\left(\varepsilon - 2 - \omega - \frac{1-2\gamma}{2}\right) + O(1)$$
$$= -\frac{n^2}{4} + \frac{n}{4}\left(\varepsilon - \delta - \frac{3}{2}\right) + O(1)$$

The two numerators are

$$-\varepsilon Q'_n - \alpha\beta Q_n - n(n+\omega)Q_n + (\alpha\beta h + n(n+\omega)a)$$
$$= \frac{n^2(2a-1)}{2} + \frac{n\omega(2a-1)}{2} + O(1)$$

and

$$-\varepsilon P'_{n-1} - \alpha\beta P_{n-1} - (n-1)(n-1+\omega)P_{n-1}$$
$$= \varepsilon\frac{n}{4} - (n^2 + n(\omega - 2) + O(1))\left(-\frac{1}{4} + \frac{1-2\gamma}{8n} + O(1/n^2)\right) + O(1)$$
$$= \frac{n^2}{4} + n\frac{2\varepsilon + 4\gamma + 2\delta - 7}{8} + O(1)$$

These estimates imply

$$\begin{cases} E_n = -\left(4a - 2 + \dfrac{2\omega(2a-1)}{n} + O(1/n^2)\right)\left(1 + \dfrac{1}{n}\left(\varepsilon - \delta - \dfrac{3}{2}\right) + O(1/n^2)\right) \\[3mm] F_n = -\left(1 + \dfrac{2\varepsilon + 4\gamma + 2\delta - 7}{2n} + O(1/n^2)\right)\left(1 + \dfrac{1}{n}\left(\varepsilon - \delta - \dfrac{3}{2}\right) + O(1/n^2)\right) \end{cases}$$

which simplifies to

$$\begin{cases} E_n = 2 - 4a + \dfrac{1-2a}{n}(2\varepsilon + 2\gamma - 5) + O(1/n^2) \\[3mm] F_n = -1 - \dfrac{4\varepsilon + 4\gamma - 10}{2n} + O(1/n^2) \end{cases}$$

and the lemma is proved. □

The asymptotic behavior of the coefficients c_n is given by the following lemma:

Lemma 8.3. *The limit*

$$\lim_{n\to\infty}\frac{c_{n+1}}{c_n}=t=\frac{\left(1-a^{-1}\right)^{1/2}+1}{\left(1-a^{-1}\right)^{1/2}-1}$$

exists, where the branch of the square root satisfies $\mathrm{Re}\left(1-a^{-1}\right)^{1/2}>0$ *(under the assumption* $|a|>1$*), and* $|t|>1$*. In the exceptional case, the limit is*

$$\lim_{n\to\infty}\frac{c_{n+1}}{c_n}=t=\frac{\left(1-a^{-1}\right)^{1/2}-1}{\left(1-a^{-1}\right)^{1/2}+1}$$

and $|t|<1$*.*

Moreover, if we write

$$\frac{c_{n+1}}{c_n}=t+\frac{\tau}{n}+O(1/n^2),\quad\text{as }n\to\infty$$

then τ *satisfies*

$$\tau=t(2\gamma+2\varepsilon-5)\frac{(1-2a)t-1}{t^2-1}$$

Proof. From Lemma 8.2 we get

$$c_{n+1}=E_nc_n+F_nc_{n-1}\tag{8.15}$$

where

$$\begin{cases}E_n=-\dfrac{\varepsilon Q'_n+\alpha\beta Q_n+n(n+\omega)Q_n-(\alpha\beta h+n(n+\omega)a)}{\varepsilon R'_{n+1}+(\alpha\beta+(n+1)(n+1+\omega))R_{n+1}}\\[2mm]F_n=-\dfrac{\varepsilon P'_{n-1}+\alpha\beta P_{n-1}+(n-1)(n-1+\omega)P_{n-1}}{\varepsilon R'_{n+1}+(\alpha\beta+(n+1)(n+1+\omega))R_{n+1}}\end{cases}$$

with dominant contributions, see (8.14)

$$\begin{cases}E_n=2-4a+\dfrac{(1-2a)(2\gamma+2\varepsilon-5)}{n}+O(1/n^2)\\[2mm]F_n=-1-\dfrac{2\gamma+2\varepsilon-5}{n}+O(1/n^2)\end{cases}\quad\text{as }n\to\infty$$

To show that c_{n+1}/c_n has a limit as $n\to\infty$, we can employ Theorem B.2 on page 183 directly. However, as in Lemma 8.1, we proceed in a somewhat different way, and write

$$a_n=t^{-n}c_n$$

where we choose t such that the new recursion relation

$$a_{n+1}=A_na_n+B_na_{n-1}$$

where

$$\begin{cases} A_n = \dfrac{E_n}{t} \\ B_n = \dfrac{F_n}{t^2} \end{cases}$$

satisfies

$$\begin{cases} A_n = 1 + \alpha + O(1/n) \\ B_n = -\alpha + O(1/n) \end{cases}$$

with $|\alpha| < 1$. From the asymptotic behavior of E_n and F_n, t has to satisfy (remember we require $A_n + B_n \to 1$ as $n \to \infty$)

$$t^2 - 2t(1 - 2a) + 1 = 0 \tag{8.16}$$

with roots

$$t_{1,2} = 1 - 2a \pm 2 \left(a^2 - a \right)^{1/2} = - \left(a^{1/2} \mp (a-1)^{1/2} \right)^2$$

which we write as

$$t_1 = \frac{A-1}{A+1}, \qquad t_2 = \frac{1}{t_1} = \frac{A+1}{A-1}$$

where

$$A = \left(1 - a^{-1} \right)^{1/2}$$

The first root satisfies

$$|t_1| = \left| \frac{A-1}{A+1} \right| < 1$$

since $A = \left(1 - a^{-1} \right)^{1/2}$ is located in the right-hand side of the complex A-plane ($\operatorname{Re} A > 0$, $|a| > 1$). Therefore, $|t_2| > 1$, and in order to satisfy the condition $|\alpha| = |t|^{-2} < 1$, we have to pick the root $t = t_2$. We can now apply Corollary B.1 on page 193, which shows that the quotient a_{n+1}/a_n converges to 1 (except in the exceptional case), which implies

$$\lim_{n \to \infty} \frac{c_{n+1}}{c_n} = t_2$$

or in the exceptional case to

$$\lim_{n \to \infty} \frac{c_{n+1}}{c_n} = t_1$$

The first part of the lemma is then proved.

To prove the conditions on τ, we once again make use of the recursion relation in (8.15) and divide by c_n. The asymptotic behavior of the coefficients then implies

$$t + \frac{\tau}{n} + O(1/n^2) = 2 - 4a + \frac{(1-2a)(2\gamma+2\varepsilon-5)}{n} + O(1/n^2)$$
$$- \left(1 + \frac{2\gamma+2\varepsilon-5}{n} + O(1/n^2) \right) \frac{1}{t + \frac{\tau}{n-1} + O(1/n^2)}$$

Use (8.16), and keep terms up to order $1/n$. We get

$$\frac{\tau}{n} = \frac{(1-2a)(2\gamma+2\varepsilon-5)}{n} + \frac{\tau}{t^2(n-1)} - \frac{2\gamma+2\varepsilon-5}{tn} + O(1/n^2)$$

Multiply with n and take the limit $n \to \infty$. This gives

$$\tau = (1-2a)(2\gamma+2\varepsilon-5) + \frac{\tau}{t^2} - \frac{2\gamma+2\varepsilon-5}{t}$$

which we solve for τ. The result is

$$\tau = t(2\gamma+2\varepsilon-5)\frac{(1-2a)t-1}{t^2-1}$$

which proves the lemma.   □

Lemma 8.3 proves that there are two possible limits t of the quotient c_{n+1}/c_n, viz., t_1 and t_2. Since $|t_1| < |t_2|$, t_2 is the limit in the general case, but under certain conditions, analogous to the result in Section 8.2, the quotient c_{n+1}/c_n converges to the other root t_1, see also (B.4) in Appendix B.

8.4.3 Domain of convergence

The main result about the domain of convergence of the series in hypergeometric polynomials in (8.8) is now collected in a theorem.

Theorem 8.2. *In the general case, when t_2 is the appropriate root in Lemma 8.3, the hypergeometric polynomial series in (8.8) converges nowhere outside the segment $[0, 1]$.*

However, when t_1 is the appropriate root in Lemma 8.3, the hypergeometric polynomial series in (8.8) converges inside the ellipse with foci at $z = 0$ and $z = 1$, passing through the point $z = a$, see Figure 8.3, with possible exception of the line connecting the two foci. Moreover, the series then converges absolutely on the ellipse, provided $\mathrm{Re}\,\varepsilon < 1$.

Proof. By the ratio test, absolute convergence of the series in (8.8) is guaranteed, provided

$$\lim_{k \to \infty} \left| \frac{c_{k+1} y_{k+1}(z)}{c_k y_k(z)} \right| = |t_n s_2(z)| < 1, \quad n = 1, 2$$

Similarly, the series diverges if $|t_n s_2(z)| > 1$, $n = 1, 2$. Lemmas 8.1 and 8.3 show that

$$s_2(z) = \frac{Z+1}{Z-1}, \qquad \begin{cases} t_1 = \dfrac{A-1}{A+1} \\[2mm] t_2 = \dfrac{A+1}{A-1} \end{cases}$$

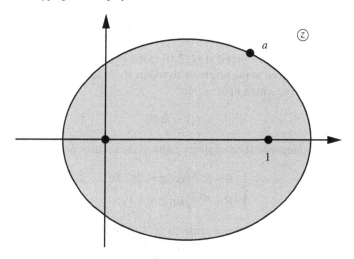

Fig. 8.3 The domain of convergence of the hypergeometric polynomial series expansion of the solution to Heun's differential equation that is analytic at the origin in the exceptional case.

where

$$\begin{cases} Z = \left(1 - z^{-1}\right)^{1/2}, & z \notin [0,1], \quad \mathrm{Re}\,Z > 0 \\ A = \left(1 - a^{-1}\right)^{1/2}, & |a| > 1, \quad \mathrm{Re}\,A > 0 \end{cases}$$

The interval $z \in [0,1]$ satisfies $\mathrm{Re}\,Z = 0$, which is the boundary of the set $\mathrm{Re}\,Z > 0$.

The result of Lemma 8.3 shows that, in general, t_2 is the proper root. The boundary of the domain of convergence, $|t_2 s_2(z)| = 1$, consists of all complex numbers z satisfying

$$\left|\frac{Z+1}{Z-1}\right| = |s_2(z)| = \frac{1}{|t_2|} = \left|\frac{A-1}{A+1}\right| < 1$$

where the last inequality follows from the requirement that $\mathrm{Re}\,A > 0$. The solution set is void, since $|Z+1| < |Z-1|$ implies that $\mathrm{Re}\,Z < 0$, which is a contradiction to the assumption $\mathrm{Re}\,Z > 0$.

The other root, t_1, gives a domain of convergence determined by

$$\left|\frac{Z+1}{Z-1}\right| = |s_2(z)| < \frac{1}{|t_1|} = \left|\frac{A+1}{A-1}\right|$$

which defines the interior of an ellipse in the complex z-plane, with foci at $z = 0, 1$ and passing through $z = a$, see Lemma D.1 in Appendix D on page 205.

A somewhat easier way of realizing that the curve is an ellipse is to fix the modulus R of $s_2(z)$, i.e.,

$$|s_2(z)| = R = \frac{1}{|t_1|} > 1$$

The equation, see (8.13),

$$s_2(z) + \frac{1}{s_2(z)} = 2 - 4z$$

then represents an ellipse centered at $(1/2, 0)$ in the complex z-plane as $s_2(z)$ varies along the circle, centered at the origin, with radius R. To see this, let $z = x + iy$, and introduce a polar representation of $s_2(z)$,

$$s_2(z) = Re^{i\phi}$$

We get, taking the real and the imaginary parts of the relation above,

$$\begin{cases} (R + R^{-1})\cos\phi = 2 - 4x \\ (R - R^{-1})\sin\phi = -4y \end{cases}$$

This equation represents an ellipse centered at $(1/2, 0)$ in the complex z-plane with the half-axes being $(R + R^{-1})/4$ and $(R - R^{-1})/4$, respectively, since $R = 1/|t_1| > 1$. The distance between the foci is $\left((R + R^{-1})^2 - (R - R^{-1})^2\right)^{1/2}/2 = 1$, so the foci are located at $z = 0, 1$.

On the ellipse, $t_1 s_2(z) = 1$, Raabe's test guarantees absolute convergence, if there exists a positive number c, such that

$$\lim_{n\to\infty} n\,\mathrm{Re}\left(\frac{c_{n+1}y_{n+1}(z)}{c_n y_n(z)} - 1\right) = -1 - c$$

In our case we have from Lemmas 8.1 and 8.3 $(t_1 s_2(z) = 1)$

$$\lim_{n\to\infty} n\,\mathrm{Re}\left(\frac{c_{n+1}y_{n+1}(z)}{c_n y_n(z)} - 1\right) = \lim_{n\to\infty} n\,\mathrm{Re}\left(\left(t_1 + \frac{\tau}{n}\right)\left(s_2(z) + \frac{\sigma(z)}{n}\right) - 1\right)$$

$$= \mathrm{Re}\left(t_1\sigma(z) + \tau s_2(z)\right)$$

$$= \mathrm{Re}\left(\frac{1 - 2\gamma}{2} + (2\gamma + 2\varepsilon - 5)\frac{(1 - 2a)t_1 - 1}{t_1^2 - 1}\right)$$

$$= \mathrm{Re}\left(\frac{1 - 2\gamma}{2} + (2\gamma + 2\varepsilon - 5)\frac{1}{2}\right) = \mathrm{Re}\,\varepsilon - 2$$

where we also used $t_1^2 = t_1(2 - 4a) - 1$. By Raabe's test, the series converges absolutely on the ellipse if $\mathrm{Re}\,\varepsilon < 1$. $\square$

8.5 Confluent Heun's equation

In Chapter 7 we analyzed two different confluent versions of the hypergeometric differential equation. These equations were obtained by letting one of the regular singular points approach infinity, and at the same time appropriately adjusting the roots of the indicial equation. The same type of procedure is, of course, also possible for Heun's equation, but, due to a more complex equation, this can be done in several alternative ways. In this section, we restrict ourselves to just one of the many existing versions. This version has certain similarities with the confluence of the second kind described in Section 7.2.

The starting point is Heun's equation, see (8.1),

$$u''(z) + \left[\frac{\gamma}{z} + \frac{\delta}{z-1} + \frac{\alpha+\beta-\gamma-\delta+1}{z-a}\right]u'(z) + \frac{\alpha\beta(z-h)}{z(z-1)(z-a)}u(z) = 0$$

where we have used (8.2) to eliminate root ε.

The aim now is to let $a \to \infty$ and at the same time adjust the roots of the indicial equation. To accomplish this, we scale $\alpha = k_1 a^{1/2}$ and $\beta = k_2 a^{1/2}$, and let $a \to \infty$. After the limit process, the new differential equation reads

$$u''(z) + \left[\frac{\gamma}{z} + \frac{\delta}{z-1}\right]u'(z) - \frac{k_1 k_2(z-h)}{z(z-1)}u(z) = 0 \qquad (8.17)$$

This is the standard form of the singly confluent Heun's equation.

An alternative form of this equation is obtained by the use of Theorem 2.1 with $z = g(t) = t^2$. The result is

$$v''(t) + \left[\frac{2\gamma-1}{t} - \frac{2t\delta}{1-t^2}\right]v'(t) - \frac{4k_1 k_2(h-t^2)}{1-t^2}v(t) = 0 \qquad (8.18)$$

A second way of confluence is addressed in Problem 8.3.

8.6 Special examples

Several special cases of Heun's differential equation and its confluent versions play an important role in mathematical physics. We illustrate with three examples.

8.6.1 Lamé's differential equation

The first example of Heun's equation is Lamé's differential equation, which is characterized by

$$\gamma = \delta = \varepsilon = \frac{1}{2} \quad \Longrightarrow \quad \alpha + \beta = \frac{1}{2}$$

This special case gives Lamé's equation

$$u''(z) + \frac{1}{2}\left[\frac{1}{z} + \frac{1}{(z-1)} + \frac{1}{(z-a)}\right]u'(z) + \frac{\alpha\beta(z-h)}{z(z-1)(z-a)}u(z) = 0$$

8.6.2 Differential equation for spheroidal functions

The prolate spheroidal coordinates (u, w, ϕ) are defined in terms of the Cartesian coordinates (x_1, x_2, x_3) as

$$(x_1, x_2, x_3) = a(\sinh u \sin w \cos \phi, \sinh u \sin w \sin \phi, \cosh u \cos w)$$

where $u \in [0, \infty)$, $w \in [0, \pi]$, and $\phi \in [0, 2\pi)$. It is convenient to introduce

$$\begin{cases} \xi_1 = \cosh u \in [1, \infty) \\ \xi_2 = \cos w \in [-1, 1] \\ \xi_3 = \phi \in [0, 2\pi) \end{cases}$$

and in these coordinates the Helmholtz equation, see Chapter 1, separates into the following equations:

$$\begin{cases} \dfrac{d}{d\xi_1}\left((\xi_1^2 - 1)\dfrac{df_1(\xi_1)}{d\xi_1}\right) - \left[\lambda - a^2 k^2 \xi_1^2 + \dfrac{\mu^2}{\xi_1^2 - 1}\right] f_1(\xi_1) = 0 \\[2mm] \dfrac{d}{d\xi_2}\left((1 - \xi_2^2)\dfrac{df_2(\xi_2)}{d\xi_2}\right) + \left[\lambda - a^2 k^2 \xi_2^2 - \dfrac{\mu^2}{1 - \xi_2^2}\right] f_2(\xi_2) = 0 \\[2mm] \dfrac{d^2 f_3(\xi_3)}{d\xi_3^2} + \mu^2 f_3(\xi_3) = 0 \end{cases}$$

where λ and μ are constants of separation. We notice that the two first equations are, from a differential equation point of view, really the same, but with different domains of definition of the independent variable. This is the differential equation of the spheroidal functions. We adopt the notation

$$(1 - x^2)f''(x) - 2xf'(x) + \left[\lambda + \chi^2(1 - x^2) - \dfrac{\mu^2}{1 - x^2}\right] f(x) = 0$$

From the separation of variables above, we identify two different domains of interest for the independent variable:

$$\begin{cases} 1) & x \in [1, \infty) \\ 2) & x \in [-1, 1] \end{cases}$$

We let $f(x) = (1 - x^2)^{\mu/2} g(x)$. The function $g(x)$ satisfies (use Theorem 2.1)

$$(1 - x^2)g''(x) - 2(\mu + 1)xg'(x) + \left[\lambda - \mu(\mu + 1) + \chi^2(1 - x^2)\right] g(x) = 0 \quad (8.19)$$

and the roots of the indicial equation are 0 and $-\mu$ for both $x = \pm 1$. Compare this equation with the confluent Heun's differential equation (8.18). This is the same equation with the constants

$$\begin{cases} \gamma = \dfrac{1}{2} \\ \delta = \mu + 1 \end{cases} \qquad \begin{cases} k_1 = -k_2 = \dfrac{\chi}{2} \\ h = \dfrac{\lambda + \chi^2 - \mu(\mu+1)}{\chi^2} \end{cases}$$

Thus, the spheroidal differential equation is a special case of the confluent Heun's differential equation.

8.6.3 Mathieu's differential equation

Mathieu's[4] differential equation is a special case of (8.19). Let $\mu = -1/2$, and (8.19) becomes

$$(1-x^2)g''(x) - xg'(x) + \left[\frac{1}{4} + \lambda + \chi^2(1-x^2)\right]g(x) = 0$$

Introduce $x = \cos\theta$ and

$$u(\theta) = g(\cos\theta)$$

We get, see Theorem 2.1,

$$u''(\theta) + \left[\frac{1}{4} + \lambda + \chi^2 \sin^2\theta\right]u(\theta) = 0$$

Compare this equation with the alternative forms of the Mathieu's differential equation found in the literature

$$\begin{cases} u''(\theta) + [a - 2q\cos 2\theta]\,u(\theta) = 0, & \text{HMF [1]} \\ u''(\theta) + [b - h^2\cos^2\theta]\,u(\theta) = 0, & \text{Morse and Feshbach [18]} \\ u''(\theta) + [a + 16q\cos 2\theta]\,u(\theta) = 0, & \text{Whittaker and Watson [31]} \end{cases}$$

We see that they all express the same equation, but with different constants.

Problems

8.1. [†]Prove that the power series of Heun's differential equation in (8.3)

$$u(z) = \sum_{n=0}^{\infty} a_n z^n$$

---

[4] Émile Léonard Mathieu (1835–1890), French mathematician.

converges absolutely on the unit circle, provided $\mathrm{Re}\,\delta < 1$ or $\mathrm{Re}\,\varepsilon < 1$, depending whether the root $t_2 = 1$ or $t_1 = 1/a$, respectively, is used.

8.2. Verify the recursion relations in (8.10) and (8.11) using the result in Section 5.8.6 and in Problem 5.7.

8.3. In Section 8.5, one type of confluence of Heun's equation was demonstrated. This type of confluence resembles the confluence of the second kind of the hypergeometric function that was demonstrated in Section 7.2. Perform a confluence of Heun's equation that parallels the confluence of the first kind of the hypergeometric function in Section 7.1.

8.4. Explicitly determine the coefficients a'_n in the power series expansion of Heun's equation in (8.3). Determine the radius of convergence of the solution, and relate the solution to the power series solution $Hl(a,h;\alpha,\beta,\gamma,\delta,\varepsilon;z)$ in Theorem 8.1.

Appendix A
The gamma function and related functions

The gamma function plays a central role in the representation of the special functions. This function and other functions derived from the gamma function are collected in this appendix together with some of their basic properties.

A.1 The gamma function $\Gamma(z)$

The gamma function is not a special function in the sense that it is a solution to a second order ordinary differential equation of the kind we have analyzed in this book, i.e., a differential equation of Fuchsian type with a finite number of regular singular points. That this is the case can be deduced from the fact that the gamma function has infinitely many poles — a property that no solution to a second order ordinary differential equation with a finite number of regular singular points can have. Instead, the gamma function has to be defined by other means. In fact, it satisfies a difference equation rather than a differential equation, see below in (A.2).

There are several ways of defining the gamma function $\Gamma(z)$. We prefer to define it by an integral representation, due to Euler, [18], viz.,

$$\Gamma(z) = \int_0^\infty e^{-t} t^{z-1} \, dt, \qquad \mathrm{Re}\, z > 0 \tag{A.1}$$

An explicit value is

$$\Gamma(1) = \int_0^\infty e^{-t} \, dt = 1$$

We easily see by integration by parts that

$$\Gamma(z+1) = z\Gamma(z), \qquad \mathrm{Re}\, z > -1, \text{ and } z \neq 0 \tag{A.2}$$

Evaluated at the non-negative integers, $\Gamma(n+1)$ coincides with the factorials, i.e.,

$$\Gamma(n+1) = n! \qquad n \in \mathbb{N}$$

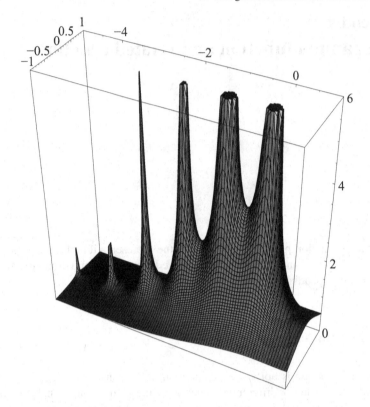

Fig. A.1 The modulus of the $\Gamma(z)$ function in the complex z-plane.

which is easy to verify by induction over n. This result shows that the gamma function is a generalization of the factorials from the non-negative integers to the complex plane.

The gamma function has simple poles at $z = 0, -1, -2, -3, \ldots$, which is seen by repeated use of (A.2), see Figure A.1,

$$\Gamma(z) = \frac{\Gamma(1+z)}{z} = \ldots = \frac{\Gamma(n+1+z)}{(n+z)(n-1+z)\ldots z} = \frac{\Gamma(n+1+z)}{(n+z)(z,n)}$$

where we introduced the Appell symbol, (α, n), defined in Section A.3.

The residue of $\Gamma(z)$ at the poles $z = 0, -1, -2, -3, \ldots$ is determined by

$$\operatorname*{Res}_{z=-n} \Gamma(z) = \lim_{z \to -n} (z+n)\Gamma(z) = \frac{\Gamma(1)}{(-n,n)} = \frac{(-1)^n}{n!}, \quad n \in \mathbb{N} \qquad (A.3)$$

since $(-n, n) = (-n)(-n+1)\ldots(-1) = (-1)^n n!$ and $\Gamma(1) = 1$.

An alternative definition of the gamma function is to use the integral representation, see Problem A.2,

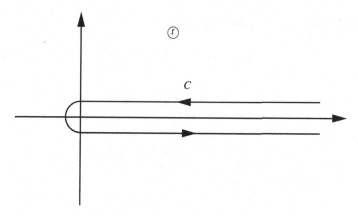

Fig. A.2 Definition of the contour C, which encircles the origin counter-clockwise.

$$\Gamma(z) = \frac{i}{2\sin \pi z} \int_C e^{-t}(-t)^{z-1} dt, \quad \forall z \in \mathbb{C}$$

where the contour C is depicted in Figure A.2.

The reciprocal of the gamma function $1/\Gamma(z)$ is an entire function with zeros at $z = 0, -1, -2, -3, \ldots$. The Weierstrass factorization theorem states that every entire function can be written as an infinite product [9]. The gamma function has the product formula

$$\frac{1}{\Gamma(z)} = ze^{\gamma z}\prod_{n=1}^{\infty}\left(1 + \frac{z}{n}\right)e^{-z/n}, \quad \forall z \in \mathbb{C}$$

where γ is the Euler–Mascheroni[1] constant defined in (A.18) on page 176.

We also frequently use

$$\Gamma(z)\Gamma(1-z) = \frac{\pi}{\sin \pi z} \tag{A.4}$$

which is proved in Problem A.1. If we evaluate this identity at $z = 1/2$, we get

$$\Gamma(1/2) = \sqrt{\pi}$$

Additional expression for products of gamma functions are, see also Problem A.5:

$$\Gamma(z)\Gamma(z+1/2) = \frac{\sqrt{\pi}}{2^{2z-1}}\Gamma(2z) \tag{A.5}$$

and

$$\Gamma(z)\Gamma(z+1/3)\Gamma(z+2/3) = \frac{2\pi}{3^{3z-1/2}}\Gamma(3z)$$

---

[1] Lorenzo Mascheroni (1750–1800), Italian mathematician.

A.2 Estimates of the gamma function

The gamma function for large arguments is determined by Stirling's[2] formula

$$\ln \Gamma(z) = \ln \sqrt{2\pi} + \left(z - \frac{1}{2}\right) \ln z - z + J(z), \quad |\arg(z)| < \pi \qquad \text{(A.6)}$$

where Binet's[3] function $J(z)$ is an analytic and bounded function in any open sector $\Gamma_\delta = \{z \in \mathbb{C} : |\arg(z)| < \pi - \delta\}$, where $\delta > 0$. Moreover, $|J(z)| \le C(\delta)/|z|$ as $z \to \infty$ in Γ_δ. The branch of the logarithm is real for real arguments. An explicit representation of Binet's function is

$$J(z) = \frac{1}{\pi} \int_0^\infty \frac{z}{z^2 + t^2} \ln \frac{1}{1 - e^{-2\pi t}} \, dt, \quad \text{Re}\, z > 0$$

From this expression of Binet's function, it is possible to analytically continue the function $J(z)$ to Γ_δ. A proof of these results is given in, e.g., [11, Sec. 8.5].

The growth rate of the gamma function is used frequently in the main text — in particular in the proofs related to the integral representations of Barnes in Chapter 5. It is convenient to summarize the results in a series of lemmas. We start with a useful estimate of the growth properties of the sine function in the complex plane.

Lemma A.1. *For each $\varepsilon \in (0, 1/2)$, there are positive constants, c_ε and C_ε, depending only on ε, but not on z, such that the sine function satisfies*

$$c_\varepsilon \le |\sin \pi z| e^{-\pi |\text{Im}\, z|} \le C_\varepsilon, \quad \text{in } \{z \in \mathbb{C} : |z + n| > \varepsilon, \forall n \in \mathbb{Z}\}$$

Proof. The modulus of the sine function in the complex plane is, $x, y \in \mathbb{R}$

$$|\sin \pi (x + iy)|^2 = \sinh^2 \pi y + \sin^2 \pi x = \frac{e^{2\pi |y|}}{4} \left\{1 + e^{-4\pi |y|} - 2e^{-2\pi |y|} \cos 2\pi x\right\}$$

For all $z = x + iy \in \mathbb{C}$, we have the estimate

$$e^{-\pi |y|} |\sin \pi (x + iy)| = \frac{1}{2} \sqrt{1 + e^{-4\pi |y|} - 2e^{-2\pi |y|} \cos 2\pi x} \le \frac{1}{2} \left(1 + e^{-2\pi |y|}\right) \le 1$$

Thus, we have proved that

$$|\sin \pi (x + iy)| \le e^{\pi |y|}, \quad \forall z \in \mathbb{C}$$

This is the right part of the inequality of the lemma. In fact, the constant $C_\varepsilon = 1$ is independent of the value of ε.

To prove the left inequality, it suffices, due to the periodicity of the sine function, i.e., $|\sin \pi(z \pm 1)| = |\sin \pi z|$, to prove the lemma in the strip $\text{Re}\, z = x \in [0, 1/2]$. We

---

[2] James Stirling (1692–1770), Scottish mathematician.
[3] Jacques Binet (1786–1856), French mathematician.

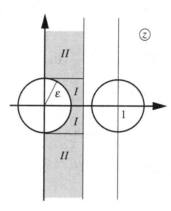

Fig. A.3 The two domains *I* and *II* in the punctuated strip *M* (shaded area) in the roof of Lemma A.1.

define for each $\varepsilon \in (0, 1/2)$ and for all complex numbers $z = x + iy$ in the punctuated strip $M = \{z \in \mathbb{C} : |z| > \varepsilon, \operatorname{Re} z \in [0, 1/2]\}$

$$f(x + iy) = 1 + e^{-4\pi|y|} - 2e^{-2\pi|y|} \cos 2\pi x, \quad \text{in } M$$

Our aim is to bound this function from below in *M*. Write $f_y(x) = f(x + iy)$, and we have

$$\frac{df_y(x)}{dx} = 4\pi e^{-2\pi|y|} \sin 2\pi x \geq 0, \quad \text{in } M$$

Therefore, for each fixed y, $f_y(x)$ is a non-decreasing function as a function of x in *M*. We divide the set *M* in two parts, *I* and *II*, see Figure A.3. In the region *I*, where $|y| \leq \varepsilon$ and $x \in [\sqrt{\varepsilon^2 - y^2}, 1/2]$, we have for $\varepsilon \in (0, 1/2)$

$$1 + e^{-4\pi|y|} - 2e^{-2\pi|y|} \cos\left(2\pi\sqrt{\varepsilon^2 - y^2}\right) = f_y(\sqrt{\varepsilon^2 - y^2}) \leq f_y(x)$$

The function on the left-hand side is bounded from below when $|y| \in [0, \varepsilon]$. To see this, rewrite the left-hand side using the notation

$$a(y) = 2\pi|y|, \qquad b(y) = 2\pi\sqrt{\varepsilon^2 - y^2}, \qquad c(x) = \sqrt{4\pi^2\varepsilon^2 - x^2}$$

as

$$f_y(\sqrt{\varepsilon^2 - y^2}) = 2e^{-a(y)} \left(\int_0^{a(y)} \sinh x \, dx + \int_0^{b(y)} \sin x \, dx \right)$$

$$= 2e^{-a(y)} \left(\int_0^{a(y)} \sinh x \, dx + \int_{a(y)}^{2\pi\varepsilon} x \frac{\sin c(x)}{c(x)} \, dx \right)$$

Since

$$\sinh x \geq x \geq x \frac{\sin c(x)}{c(x)} = \frac{d}{dx} \cos c(x), \quad x \geq 0$$

we get ($|y| \in [0, \varepsilon]$)

$$f_y\left(\sqrt{\varepsilon^2 - y^2}\right) \geq 2e^{-a(y)} \int_0^{2\pi\varepsilon} x \frac{\sin c(x)}{c(x)} \, dx$$

$$\geq 2e^{-2\pi\varepsilon} \int_0^{2\pi\varepsilon} \frac{d}{dx} \cos c(x) \, dx = 2e^{-2\pi\varepsilon} \left(1 - \cos 2\pi\varepsilon\right)$$

Thus, in region I, for each $\varepsilon \in (0, 1/2)$, we have

$$2e^{-2\pi\varepsilon} \left(1 - \cos 2\pi\varepsilon\right) = \left(2e^{-\pi\varepsilon} \sin \pi\varepsilon\right)^2 \leq f_y(x)$$

In region II, $|y| \geq \varepsilon$, and we have

$$\left(1 - e^{-2\pi\varepsilon}\right)^2 \leq \left(1 - e^{-2\pi|y|}\right)^2 = 1 + e^{-4\pi|y|} - 2e^{-2\pi|y|} = f_y(0) \leq f_y(x)$$

To summarize, for each $\varepsilon \in (0, 1/2)$ and for all $z = x + iy \in M$, we have the estimate

$$4c_\varepsilon^2 \leq f_y(x) = 4e^{-2\pi|y|} |\sin \pi(x + iy)|^2$$

where

$$c_\varepsilon = \min\left\{e^{-\pi\varepsilon} \sinh \pi\varepsilon, e^{-\pi\varepsilon} \sin \pi\varepsilon\right\} = e^{-\pi\varepsilon} \sin \pi\varepsilon$$

Thus, we have proved that there exists a constant c_ε, such that

$$c_\varepsilon e^{\pi|\operatorname{Im} z|} \leq |\sin \pi z|, \quad z \in M$$

which by periodicity holds for all $\{z \in \mathbb{C} : |z + n| > \varepsilon, \forall n \in \mathbb{Z}\}$, and the lemma is proved. $\square$

The argument of the sine function can also be shifted leading to the following corollary:

Corollary A.1. *For each $\varepsilon \in (0, 1/2)$ and $\alpha \in \mathbb{C}$, there are positive constants, $c_{\varepsilon,\alpha}$ and $C_{\varepsilon,\alpha}$, depending only on ε and α, but not on z, such that the sine function satisfies*

$$c_{\varepsilon,\alpha} \leq |\sin \pi(z + \alpha)| e^{-\pi|\operatorname{Im} z|} \leq C_{\varepsilon,\alpha}, \text{ in } \{z \in \mathbb{C} : |z + \alpha + n| > \varepsilon, \forall n \in \mathbb{Z}\}$$

Proof. Lemma A.1 applied with $z + \alpha$ reads

$$c_\varepsilon \leq |\sin \pi(z + \alpha)| e^{-\pi|\operatorname{Im}(z+\alpha)|} \leq C_\varepsilon, \text{ in } \{z \in \mathbb{C} : |z + \alpha + n| > \varepsilon, \forall n \in \mathbb{Z}\}$$

The lemma follows from the triangle inequality

$$|\operatorname{Im} z| - |\operatorname{Im} \alpha| \leq |\operatorname{Im}(z + \alpha)| \leq |\operatorname{Im} z| + |\operatorname{Im} \alpha|$$

which leads to

$$c_{\varepsilon,\alpha} = e^{-\pi|\operatorname{Im} \alpha|} c_\varepsilon$$

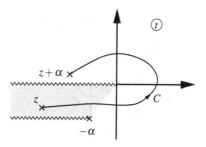

Fig. A.4 The contour C in the complex plane that connects the two points z and $z + \alpha$, assuming $\operatorname{Im} \alpha > 0$. The shaded area denotes the region between the two cuts.

and

$$C_{\varepsilon,\alpha} = e^{\pi |\operatorname{Im} \alpha|} C_\varepsilon$$

and the proof is completed. $\quad \square$

We have also need for the following useful result:

Lemma A.2. *For arbitrary* $\alpha, \gamma \in \mathbb{C}$, *the **real part** of the function*

$$f(z) = (z + \gamma)\left(\ln(z + \alpha) - \ln z\right)$$

is a bounded function in $\Gamma_{\varepsilon,\alpha} = \{z \in \mathbb{C} : |z + \alpha| \geq \varepsilon, |z| \geq \varepsilon\}$, *where* $\varepsilon > 0$, *and where the logarithms belong to the principal branch.*[4] *Explicitly, there exist constants* $c_{\varepsilon,\alpha,\gamma}$ *and* $C_{\varepsilon,\alpha,\gamma}$ *that depend only on* ε, α, *and* γ, *but not on* z, *such that*

$$c_{\varepsilon,\alpha,\gamma} \leq \operatorname{Re} f(z) \leq C_{\varepsilon,\alpha,\gamma}, \quad \text{in } \Gamma_{\varepsilon,\alpha} \tag{A.7}$$

Proof. We start with the identity

$$\ln(z + \alpha) - \ln z = \int_C \frac{dt}{t}, \quad z \in \Gamma_{\varepsilon,\alpha}$$

where C is a contour in the complex t-plane that connects the two points z and $z + \alpha$ without crossing the two branch cuts, $|\arg(z)| = \pi$ and $|\arg(z + \alpha)| = \pi$, see Figure A.4.

If the two points, z and $z + \alpha$, lie on the same side of the two cuts, i.e., both z and $z + \alpha$ are located outside the shaded area in Figure A.4, we can connect the points with a straight line, and the real part of the function $f(z)$ can be estimated for large

---

[4] The principal branch of the logarithm is defined as

$$\ln z = \ln |z| + i \arg(z), \quad \arg(z) \in (-\pi, \pi]$$

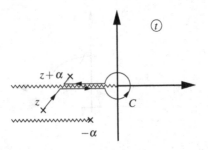

Fig. A.5 The deformed contour C in the complex plane that connects the two points z and $z + \alpha$, assuming Im $\alpha > 0$. Notice that the contour C ends on the principal branch.

arguments as[5]

$$|\mathrm{Re}\, f(z)| = \left|\mathrm{Re} \int_0^1 \frac{(z+\gamma)\alpha\,dt}{z+t\alpha}\right| \leq |\alpha| \int_0^1 \frac{|1+\gamma/z|}{|1+t\alpha/z|}\,dt \leq |\alpha|\left(1+O(|z|^{-1})\right)$$

which is bounded in $\Gamma_{\varepsilon,\alpha}$.

However, if the two points, z and $z + \alpha$, are located on opposite sides of a branch cut, i.e., either z or $z + \alpha$ is located in the shaded area in Figure A.4, we have to compensate with a circle around the branch point, see Figure A.5, in order to end the contour C on the principal branch. The real part of the function $f(z)$ then is

$$\mathrm{Re}\, f(z) = \mathrm{Re}\left\{(z+\gamma)\left(\int_0^1 \frac{\alpha\,dt}{z+t\alpha} \pm 2\pi i\right)\right\}, \quad z \in \Gamma_{\varepsilon,\alpha}$$

where the upper or lower sign in the last term depends on whether Im $\alpha > 0$ or Im $\alpha < 0$, respectively. The real part of the extra term, however, is always bounded, since Im z (in contrast to Re z) is bounded in this case, and the lemma is proved. □

The next lemma estimates the growth properties of the gamma function from both above and below.

Lemma A.3. *For each $\varepsilon \in (0, 1/2)$, there are positive constants, c_ε and C_ε, depending only on ε, but not on z, such that the gamma function satisfies*[6]

$$c_\varepsilon \leq |\Gamma(z)|\, e^{-\mathrm{Re}(z-\frac{1}{2})\ln|z| + \mathrm{Im}\,z\,\arg(z) + \mathrm{Re}\,z} \leq C_\varepsilon, \ \text{in } \{z \in \mathbb{C} : |z+n| > \varepsilon, \forall n \in \mathbb{N}\}$$

Proof. We first prove the statement in the lemma in the right-hand z-plane, or more precisely, for all $z \in \{z \in \mathbb{C} : \mathrm{Re}\, z \geq 0 \text{ and } |z| > \varepsilon\}$. For all such z, Stirling's formula, (A.6), reads

$$\ln \Gamma(z) = \ln \sqrt{2\pi} + \left(z - \frac{1}{2}\right)\ln z - z + O(1/|z|)$$

---

[5] In fact, not only the real part of the function is bounded, but also the modulus of the function itself.

[6] Note that Im $z\,\arg(z)$ is continuous across the negative real axis.

The real part of this expression is

$$\operatorname{Re}\ln\Gamma(z) = \ln\sqrt{2\pi} + \operatorname{Re}\left(z - \frac{1}{2}\right)\ln|z| - \operatorname{Im} z \arg(z) - \operatorname{Re} z + O(1/|z|)$$

and, therefore, for each $\varepsilon > 0$, there exist constants $C_\varepsilon > c_\varepsilon > 0$, depending only on ε, but not on z, such that

$$c_\varepsilon \le |\Gamma(z)| e^{-\operatorname{Re}\left(z - \frac{1}{2}\right)\ln|z| + \operatorname{Im} z \arg(z) + \operatorname{Re} z} \le C_\varepsilon$$

for all $z \in \{z \in \mathbb{C} : \operatorname{Re} z \ge 0 \text{ and } |z| > \varepsilon\}$. This proves the estimate in the right-hand complex z-plane. In fact, this estimate holds for all z such that $|z| > \varepsilon$ and $|\arg(z)| < \pi - \delta$, where $\delta > 0$. However, we also want to estimate the gamma function on the negative real axis away from the singular points at the non-positive integers.

We now focus on an estimate in the left-hand z-plane, more precisely, for an $\varepsilon \in (0, 1/2)$, and for $z \in M_\varepsilon = \{z \in \mathbb{C} : |z + n| > \varepsilon, \forall n \in \mathbb{N}\} \cap \{z \in \mathbb{C} : \operatorname{Re} z < 0\}$. We start by utilizing (A.4), i.e.,

$$\Gamma(z) = \frac{\pi}{\Gamma(1 - z)\sin\pi z} \tag{A.8}$$

Note that $\operatorname{Re}(1 - z) > 1$ when $z \in M_\varepsilon$, and therefore, by the result above, there exists constants $C_1 > c_1 > 0$, depending only on ε, but not on z, such that

$$c_1 \le \frac{1}{|\Gamma(1 - z)|} e^{\operatorname{Re}\left(\frac{1}{2} - z\right)\ln|1 - z| + \operatorname{Im} z \arg(1 - z) - \operatorname{Re}(1 - z)} \le C_1, \quad z \in M_\varepsilon$$

Using Lemma A.1 we can estimate (A.8) in M_ε. There exist constants $C_2 > c_2 > 0$, depending only on ε, but not on z, such that

$$c_2 \le |\Gamma(z)| e^{\operatorname{Re}\left(\frac{1}{2} - z\right)\ln|1 - z| + \operatorname{Im} z \arg(1 - z) + |\operatorname{Im} z|\pi + \operatorname{Re} z} \le C_2, \text{ in } M_\varepsilon$$

As a consequence of Lemma A.2, see (A.7), there exist constants C_3 and c_3, depending only on ε, but not on z, such that

$$c_3 \le \underbrace{\operatorname{Re}\left(\frac{1}{2} - z\right)(\ln|1 - z| - \ln|z|) - \operatorname{Im}(-z)(\arg(1 - z) - \arg(-z))}_{\operatorname{Re}\left\{\left(\frac{1}{2} - z\right)(\ln(1 - z) - \ln(-z))\right\}} \le C_3$$

and we get

$$c_\varepsilon \le |\Gamma(z)| e^{-\operatorname{Re}\left(z - \frac{1}{2}\right)\ln|z| + \operatorname{Im} z \arg(-z) + |\operatorname{Im} z|\pi + \operatorname{Re} z} \le C_\varepsilon$$

or

$$c_\varepsilon \le |\Gamma(z)| e^{-\operatorname{Re}\left(z - \frac{1}{2}\right)\ln|z| + \operatorname{Im} z \arg(z) + \operatorname{Re} z} \le C_\varepsilon$$

since

$$\arg(-z) = \arg(z) - \begin{cases} \pi, & \mathrm{Im}\, z \geq 0 \\ -\pi, & \mathrm{Im}\, z \leq 0 \end{cases} \quad \text{in } M_\varepsilon$$

The combination of the results in the right- and the left-hand z-planes then proves the lemma. □

It is also possible to bound the shifted gamma function.

Corollary A.2. *For each $\alpha \in \mathbb{C}$ and each $\varepsilon \in (0, 1/2)$, there are positive constants, $c_{\varepsilon,\alpha}$ and $C_{\varepsilon,\alpha}$, depending only on ε and α, but not on z, such that the gamma function satisfies*

$$c_{\varepsilon,\alpha} \leq |\Gamma(z+\alpha)| e^{-\mathrm{Re}\left(z+\alpha-\frac{1}{2}\right)\ln|z| + \mathrm{Im}\, z \arg(z) + \mathrm{Re}\, z} \leq C_{\varepsilon,\alpha}, \; \text{in } M_{\varepsilon,\alpha}$$

where $M_{\varepsilon,\alpha} = \{z \in \mathbb{C} : |z+\alpha+n| > \varepsilon, |z| > \varepsilon, \forall n \in \mathbb{N}\}$ and $|\arg(z)| \leq \pi$.

Proof. From Lemma A.3 we have

$$c_\varepsilon \leq |\Gamma(z+\alpha)| e^{-\mathrm{Re}\left(z+\alpha-\frac{1}{2}\right)\ln|z+\alpha| + \mathrm{Im}(z+\alpha)\arg(z+\alpha) + \mathrm{Re}(z+\alpha)} \leq C_\varepsilon, \quad z \in M_{\varepsilon,\alpha}$$

Lemma A.2, see (A.7), implies that there exist constants c_1 and C_1, depending only on ε and α, but not on z, such that

$$c_1 \leq \underbrace{\mathrm{Re}\left(z+\alpha-\frac{1}{2}\right)(\ln|z+\alpha| - \ln|z|) - \mathrm{Im}(z+\alpha)(\arg(z+\alpha) - \arg(z))}_{\mathrm{Re}\left\{\left(z+\alpha-\frac{1}{2}\right)(\ln(z+\alpha)-\ln z)\right\}} \leq C_1$$

We get

$$c_{\varepsilon,\alpha} \leq |\Gamma(z+\alpha)| e^{-\mathrm{Re}\left(z+\alpha-\frac{1}{2}\right)\ln|z| + \mathrm{Im}\, z \arg(z) + \mathrm{Re}\, z} \leq C_{\varepsilon,\alpha}, \quad z \in M_{\varepsilon,\alpha}$$

and the corollary is proved. □

The quotient between two gamma functions are frequently used in the text. The next lemma summarizes the growth properties.

Lemma A.4. *Define for each $\varepsilon \in (0, 1/2)$ and $\alpha \in \mathbb{C}$ the set, see Figure A.6,*

$$\mathbb{C}_{\varepsilon,\alpha} = \{z \in \mathbb{C} : |z+\alpha+n| > \varepsilon \text{ and } |z+n| > \varepsilon, \forall n \in \mathbb{N}\}$$

Then for any $\alpha \in \mathbb{C}$, and all $z \in \mathbb{C}_{\varepsilon,\alpha}$

$$c_{\varepsilon,\alpha} e^{\mathrm{Re}\, \alpha \ln|z|} \leq \left| \frac{\Gamma(\alpha+z)}{\Gamma(z)} \right| \leq C_{\varepsilon,\alpha} e^{\mathrm{Re}\, \alpha \ln|z|}$$

where $C_{\varepsilon,\alpha} > c_{\varepsilon,\alpha} > 0$ are constants that depend on ε and α, but not on z.

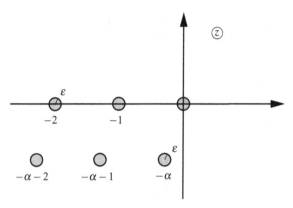

Fig. A.6 The set $\mathbb{C}_{\varepsilon,\alpha}$ (unshaded area) in the complex z-plane.

Proof. Lemma A.3 and Corollary A.2 imply that there exist constants, such that

$$c_1 \le |\Gamma(z)| \, e^{-\operatorname{Re}\left(z-\frac{1}{2}\right)\ln|z| + \operatorname{Im} z \arg(z) + \operatorname{Re} z} \le C_1, \text{ in } \mathbb{C}_{\varepsilon,\alpha}$$

and

$$c_2 \le |\Gamma(z+\alpha)| \, e^{-\operatorname{Re}\left(z+\alpha-\frac{1}{2}\right)\ln|z| + \operatorname{Im} z \arg(z) + \operatorname{Re} z} \le C_2, \text{ in } \mathbb{C}_{\varepsilon,\alpha}$$

Then for all $z \in \mathbb{C}_{\varepsilon,\alpha}$, we have

$$c_{\varepsilon,\alpha} \le \left|\frac{\Gamma(\alpha+z)}{\Gamma(z)}\right| e^{-\operatorname{Re}\alpha\ln|z|} \le C_{\varepsilon,\alpha}, \text{ in } \mathbb{C}_{\varepsilon,\alpha}$$

and the proof of the lemma is completed. $\square$

A.3 The Appell symbol

Related to the gamma function is the rising factorial, also known as the Appell symbol or Pochhammer symbol, (α, n), defined for non-negative integers n as [4]

$$(\alpha,n) = \prod_{v=0}^{n-1}(v+\alpha) = \alpha(\alpha+1)(\alpha+2)\ldots(\alpha+n-1) = \frac{\Gamma(\alpha+n)}{\Gamma(\alpha)}, \qquad (\alpha,0) = 1$$

As a consequence of this definition, we easily derive the following simple recursion relations:

$$\begin{cases} (\alpha,n) = \alpha(\alpha+1,n-1) \\ (\alpha,n)(\alpha+n) = (\alpha,n+1) = \alpha(\alpha+1,n) \end{cases} \qquad (A.9)$$

The Appell symbol for negative integers n can be defined by repeated use of (A.4). We have for non-integer values of α

$$
\begin{aligned}
\Gamma(\alpha - n) &= \frac{1}{\Gamma(1-\alpha+n)} \frac{\pi}{\sin\pi(\alpha-n)} = \frac{(-1)^n}{\Gamma(1-\alpha+n)} \frac{\pi}{\sin\pi\alpha} \\
&= \frac{(-1)^n\Gamma(\alpha)\Gamma(1-\alpha)}{\Gamma(1-\alpha+n)} = \frac{(-1)^n\Gamma(\alpha)}{(1-\alpha,n)}
\end{aligned}
\tag{A.10}
$$

from which we define

$$
(\alpha,-n) = \frac{\Gamma(\alpha-n)}{\Gamma(\alpha)} = \frac{(-1)^n}{(1-\alpha,n)}
\tag{A.11}
$$

Similarly,

$$
(-n,n) = (-n)(-n+1)(-n+2)\ldots(-1) = (-1)^n n!
\tag{A.12}
$$

In the main text the growth rate of the Appell symbol is often analyzed. The following lemma is then useful:

Lemma A.5. *For any $\alpha \in \mathbb{C}$, the Appell symbol satisfies*

$$
\frac{(\alpha,n)}{n!} = \frac{n^{\alpha-1}}{\Gamma(\alpha)}(1+O(1/n))
$$

Proof. We investigate the quotient

$$
\ln\frac{(\alpha,n)\Gamma(\alpha)}{n!\,n^{\alpha-1}} = \ln\frac{\Gamma(\alpha+n)}{n!\,n^{\alpha-1}} = \ln\Gamma(\alpha+n) - \ln\Gamma(n+1) - (\alpha-1)\ln n
$$

as $n \to \infty$. Stirling's formula, (A.6), implies for integer values of n

$$
\begin{aligned}
\ln\frac{(\alpha,n)\Gamma(\alpha)}{n!\,n^{\alpha-1}} &= \left(\alpha+n-\frac{1}{2}\right)\ln(\alpha+n) - \left(n+\frac{1}{2}\right)\ln(n+1) \\
&\quad - (\alpha-1)(1+\ln n) + O(1/n) \\
&= \left(\alpha+n-\frac{1}{2}\right)\ln\frac{\alpha+n}{n+1} - (\alpha-1)\left(1-\ln\frac{n+1}{n}\right) + O(1/n) \\
&= \left(\alpha+n+1-\frac{3}{2}\right)\ln\left(1+\frac{\alpha-1}{n+1}\right) - (\alpha-1) + O(1/n) \\
&= O(1/n)
\end{aligned}
$$

since $\ln(1+z) = z + O(z^2)$. Therefore,

$$
\ln\frac{(\alpha,n)}{n!\,n^{\alpha-1}} = -\ln\Gamma(\alpha) + O(1/n)
$$

or

$$
\frac{(\alpha,n)}{n!\,n^{\alpha-1}} = \frac{1}{\Gamma(\alpha)}(1+O(1/n))
$$

since

$$e^{O(1/n)} = 1 + O(1/n)$$

and the lemma is proved.   □

A.4 Psi (digamma) function

The logarithmic derivative of the gamma function is often used, and it is called the psi (digamma) function.

$$\psi(z) = \frac{d}{dz} \ln \Gamma(z) = \frac{\Gamma'(z)}{\Gamma(z)}$$

A relation similar to the relation (A.4) also holds for the psi function. We have, see Problem A.4,

$$\psi(1-z) = \psi(z) + \pi \cot \pi z \qquad (A.13)$$

The ψ function has closed form expressions for the half-integer values, i.e.,

$$\psi(n+1/2) = -\gamma - 2\ln 2 + 2\left(1 + \frac{1}{3} + \ldots + \frac{1}{2n-1}\right), \quad n \in \mathbb{Z}_+$$

where γ is the Euler–Mascheroni constant, which is defined in (A.18). For negative half integers we use (A.13), and get

$$\psi(-n+1/2) = \psi(n+1/2), \quad n \in \mathbb{Z}_+$$

A.5 Binomial coefficient

Related to the gamma function is the binomial coefficient. For non-negative integers n and k, the binomial coefficient is defined as

$$\binom{n}{k} = \frac{n!}{(n-k)!k!}$$

which we extend to all integer values k by

$$\binom{n}{k} = 0 \quad \text{if } k < 0 \text{ or } k > n$$

The binomial coefficient for a non-integer value α can be expressed in the gamma function, viz.,

$$\binom{\alpha}{n} = \frac{\Gamma(\alpha+1)}{\Gamma(\alpha+1-n)n!} \tag{A.14}$$

and, using (A.4) and $z\Gamma(z) = \Gamma(z+1)$,

$$\binom{-\alpha}{n} = \frac{\Gamma(1-\alpha)}{\Gamma(1-\alpha-n)n!} = \frac{\pi}{\sin\pi\alpha}\frac{1}{\Gamma(\alpha)\Gamma(1-\alpha-n)n!}$$

$$= (-1)^{n-1}\frac{\pi}{\sin\pi\alpha}\frac{(1+\alpha)\cdot\ldots\cdot(n-1+\alpha)}{\Gamma(\alpha)(-\alpha-1)\cdot\ldots\cdot(-\alpha-n+1)\Gamma(1-\alpha-n)n!}$$

$$= (-1)^{n-1}\frac{\pi}{\sin\pi\alpha}\frac{(1+\alpha)(2+\alpha)\cdot\ldots\cdot(n-1+\alpha)}{\Gamma(\alpha)\Gamma(-\alpha)n!}$$

$$= (-1)^{n}\frac{(1+\alpha)(2+\alpha)\cdot\ldots\cdot(n-1+\alpha)\Gamma(1+\alpha)}{\Gamma(\alpha)n!}$$

$$= (-1)^{n}\frac{\Gamma(n+\alpha)}{\Gamma(\alpha)n!} = (-1)^{n}\frac{(\alpha,n)}{n!} \tag{A.15}$$

A.6 The beta function $B(x,y)$

The beta function $B(x,y)$ is defined as

$$B(x,y) = \int_0^1 t^{x-1}(1-t)^{y-1}\,dt, \qquad \mathrm{Re}\,x, \mathrm{Re}\,y > 0 \tag{A.16}$$

It can be proved that this integral is a quotient of gamma functions. We have, see Problem A.3,

$$B(x,y) = \frac{\Gamma(x)\Gamma(y)}{\Gamma(x+y)} \tag{A.17}$$

A.7 Euler–Mascheroni constant

The Euler–Mascheroni constant γ is defined as

$$\gamma = \lim_{n\to\infty}\left(1 + \frac{1}{2} + \frac{1}{3} + \ldots + \frac{1}{n} - \ln n\right) = \lim_{n\to\infty}\left(\sum_{k=1}^{n}\frac{1}{k} - \ln n\right) \approx 0.5772156649\ldots \tag{A.18}$$

This definition is equivalent to

$$\gamma = \sum_{k=1}^{\infty}\left\{\frac{1}{k} - \ln\left(1 + \frac{1}{k}\right)\right\} \tag{A.19}$$

In fact, the partial sum of the expression can easily be rewritten as

$$\sum_{k=1}^{n}\left\{\frac{1}{k}-\ln\left(1+\frac{1}{k}\right)\right\} = \sum_{k=1}^{n}\frac{1}{k}-\ln\left(\prod_{k=1}^{n}\frac{k+1}{k}\right) = \sum_{k=1}^{n}\frac{1}{k}-\ln(n+1)$$

$$= \sum_{k=1}^{n}\frac{1}{k}-\ln n+\ln\frac{n}{n+1}$$

from which the equivalence follows.

From the definition of γ in equation (A.18) and Lemma B.5 on page 193, with $f(x)=1/x$ and $m=1$, we get (sharp inequalities in this explicit example)

$$\frac{1}{n} < \sum_{k=1}^{n}\frac{1}{k}-\int_{1}^{n}\frac{dx}{x} = \sum_{k=1}^{n}\frac{1}{k}-\ln n < 1$$

and in the limit as $n\to\infty$, this gives $0<\gamma<1$ (the limits are not reached in the limit process). A more sharp estimate of γ is obtained in Problem A.6.

Problems

A.1. †Prove equation (A.4), i.e.,

$$\Gamma(z)\Gamma(1-z) = \frac{\pi}{\sin \pi z}$$

A.2. †Starting from (A.1), prove

$$\Gamma(z) = \frac{i}{2\sin \pi z}\int_{C}e^{-t}(-t)^{z-1}\,dt, \quad \forall z\in\mathbb{C}$$

and

$$\frac{1}{\Gamma(z)} = \frac{i}{2\pi}\int_{C}e^{-t}(-t)^{-z}\,dt, \quad \forall z\in\mathbb{C}$$

A.3. †Prove the relation between the beta function and the product of gamma functions in (A.17), i.e.,

$$B(x,y) = \frac{\Gamma(x)\Gamma(y)}{\Gamma(x+y)}$$

A.4. Show (A.13), i.e.,

$$\psi(1-z) = \psi(z)+\pi\cot\pi z$$

A.5. Show (A.5), i.e.,

$$\Gamma(z)\Gamma(z+1/2) = \frac{\sqrt{\pi}}{2^{2z-1}}\Gamma(2z)$$

A.6. †Show that the Euler–Mascheroni constant γ satisfies

$$0.998\gamma \approx 0.575804 \approx \frac{3}{2} - \frac{4}{3}\ln 2 < \gamma < \frac{8}{3} - 2\ln 2 - \frac{7}{12}\zeta(3) \approx 0.579172 \approx 1.003\gamma$$

where $\zeta(3)$ is the Riemann zeta function, see Section B.4.1 on page 194. Compare this estimate with the exact value $\gamma =\approx 0.5772156649\ldots$.

Hint: Use the inequality

$$1 + \frac{1}{12}\frac{1}{x^2 + x + 1/4} < \left(x + \frac{1}{2}\right)\ln\left(1 + \frac{1}{x}\right) < 1 + \frac{1}{12}\frac{1}{x^2 + x}, \quad x > 0$$

Appendix B
Difference equations

The asymptotic behavior of solutions to difference equations or recursion relations is the subject of this appendix. For simplicity, we restrict ourselves to second order recursion relations, which is the situation met in this textbook, and we examine the asymptotic behavior of their solutions for large index values. Some of the results are presented without proofs, and in these cases we give references to the relevant literature.

B.1 Second order recursion relations

The second order recursion relation of interest is

$$
\begin{cases}
a_{n+1} = A_n a_n + B_n a_{n-1}, & n \in \mathbb{Z}_+ \\
a_1 = A_0 a_0
\end{cases}
\tag{B.1}
$$

or, if we define $B_0 = 0$,

$$
a_{n+1} = A_n a_n + B_n a_{n-1}, \qquad n \in \mathbb{N}
$$

The following lemma shows that the coefficients a_n can be written in terms of a determinant times the first coefficient a_0.

Lemma B.1. *Let the coefficients $\{a_n\}_{n=0}^{\infty}$, for given a_0, be generated by*

$$
a_{n+1} = A_n a_n + B_n a_{n-1}, \qquad n \in \mathbb{Z}_+
$$

and initialized by

$$
a_1 = A_0 a_0
$$

Denote $D_n = a_n/a_0$. Then D_{n+1} is expressed as the determinant

$$D_{n+1} = \begin{vmatrix} A_0 & -1 & 0 & 0 & 0 & \cdots & \cdots & \cdots \\ B_1 & A_1 & -1 & 0 & 0 & \cdots & \cdots & \cdots \\ 0 & B_2 & A_2 & -1 & 0 & \cdots & \cdots & \cdots \\ \vdots & \vdots & \vdots & \vdots & \vdots & \cdots & \vdots & \vdots \\ \cdots & \cdots & \cdots & \cdots & \cdots & \cdots & A_{n-1} & -1 \\ 0 & 0 & 0 & 0 & 0 & \cdots & B_n & A_n \end{vmatrix}$$

and, thus, the determinants satisfy the same recursion relation as the coefficients a_n, i.e.,

$$D_{n+1} = A_n D_n + B_n D_{n-1}, \qquad n \in \mathbb{Z}_+$$

Proof. First let $n = 0$. The relation is then

$$A_0 = \frac{A_0 a_0}{a_0} = \frac{a_1}{a_0} = D_1$$

which proves the statement is consistent for $n = 0$.

We prove this lemma by induction over $n \in \mathbb{Z}_+$. The relation for $n = 1$ is

$$\begin{vmatrix} A_0 & -1 \\ B_1 & A_1 \end{vmatrix} = A_0 A_1 + B_1 = \frac{A_0 A_1 a_0 + B_1 a_0}{a_0} = \frac{A_1 a_1 + B_1 a_0}{a_0} = \frac{a_2}{a_0} = D_2$$

which proves the induction statement for $n = 1$. Assume the lemma is true for $k = 1, 2, \ldots, n$ and prove it for $n + 1$. We then get by expanding the determinant along the last column and using the induction assumption

$$\begin{vmatrix} A_0 & -1 & 0 & 0 & 0 & \cdots & \cdots & \cdots \\ B_1 & A_1 & -1 & 0 & 0 & \cdots & \cdots & \cdots \\ 0 & B_2 & A_2 & -1 & 0 & \cdots & \cdots & \cdots \\ \vdots & \vdots & \vdots & \vdots & \vdots & \cdots & \vdots & \vdots \\ \cdots & \cdots & \cdots & \cdots & \cdots & \cdots & A_{n-1} & -1 \\ 0 & 0 & 0 & 0 & 0 & \cdots & B_n & A_n \end{vmatrix}$$

$$= A_n D_n + \begin{vmatrix} A_0 & -1 & 0 & 0 & 0 & \cdots & \cdots & \cdots \\ B_1 & A_1 & -1 & 0 & 0 & \cdots & \cdots & \cdots \\ 0 & B_2 & A_2 & -1 & 0 & \cdots & \cdots & \cdots \\ \vdots & \vdots & \vdots & \vdots & \vdots & \cdots & \vdots & \vdots \\ \cdots & \cdots & \cdots & \cdots & \cdots & \cdots & A_{n-2} & -1 \\ 0 & 0 & 0 & 0 & 0 & \cdots & 0 & B_n \end{vmatrix} = A_n D_n + B_n D_{n-1}$$

We then have

$$\begin{vmatrix} A_0 & -1 & 0 & 0 & 0 & \cdots & \cdots & \cdots \\ B_1 & A_1 & -1 & 0 & 0 & \cdots & \cdots & \cdots \\ 0 & B_2 & A_2 & -1 & 0 & \cdots & \cdots & \cdots \\ \vdots & \vdots & \vdots & \vdots & \vdots & \cdots & & \vdots \\ \cdots & \cdots & \cdots & \cdots & \cdots & \cdots & A_{n-1} & -1 \\ 0 & 0 & 0 & 0 & 0 & \cdots & B_n & A_n \end{vmatrix} = \frac{A_n a_n + B_n a_{n-1}}{a_0} = \frac{a_{n+1}}{a_0} = D_{n+1}$$

and the lemma is proved. $\square$

This following lemma shows that the original recursion relation in (B.1), under certain conditions, can be transformed into a form where the first coefficient $A_n = 1$.

Lemma B.2. *Let the coefficients* $\{a_n\}_{n=0}^{\infty}$, *for given* a_0, *be generated by*

$$a_{n+1} = A_n a_n + B_n a_{n-1}, \qquad n \in \mathbb{Z}_+$$

and initialized by

$$a_1 = A_0 a_0$$

If $A_k \neq 0$, $k \in \mathbb{N}$, *then the sequence* $\{a_n\}_{n=0}^{\infty}$, *for given* a_0, *can be found from the recursion relation*

$$b_{n+1} = b_n + C_n b_{n-1}, \qquad n \in \mathbb{Z}_+$$

and initialized by the same value as the original recursion relation, i.e.,

$$b_0 = a_0$$

where

$$C_n = \frac{B_n}{A_n A_{n-1}}, \qquad n \in \mathbb{Z}_+$$

and

$$a_n = A_{n-1} \cdot \ldots \cdot A_1 A_0 b_n, \qquad n \in \mathbb{Z}_+$$

Proof. The statement of the lemma is easily seen if we insert $a_n = \alpha_n b_n$, assuming $\alpha_0 = 1$, i.e.,

$$\alpha_{n+1} b_{n+1} = A_n \alpha_n b_n + B_n \alpha_{n-1} b_{n-1}, \qquad n \in \mathbb{Z}_+$$

and determine α_n such that

$$\alpha_{n+1} = A_n \alpha_n \quad \Longrightarrow \quad \alpha_{n+1} = A_n A_{n-1} \cdot \ldots \cdot A_1 A_0, \quad n \in \mathbb{N}$$

The a_n coefficient then is

$$a_n = A_{n-1} \cdot \ldots \cdot A_1 A_0 b_n, \quad n \in \mathbb{Z}_+$$

We get after division of α_{n+1} (note that $\alpha_{n+1} \neq 0$)

$$b_{n+1} = b_n + \frac{B_n \alpha_{n-1}}{\alpha_{n+1}} b_{n-1} = b_n + \frac{B_n}{A_n A_{n-1}} b_{n-1}, \qquad n \in \mathbb{Z}_+$$

and the lemma is proved.   □

B.2 Poincaré–Perron theory

The convergence of the quotient a_{n+1}/a_n as $n \to \infty$ is crucial in the development of the theory presented in this book, especially when finding the radius of convergence for power series. This result is often referred to as the Poincaré–Perron theory. A comprehensive treatment of this problem is found in Ref. 6 and originates from pioneer works by Poincaré[1] and Perron.[2] In this section, we give an overview of the main results of this theory.

We immediately see that if the coefficients A_n and B_n in (B.1) satisfy

$$\begin{cases} \lim_{n \to \infty} A_n = A \\ \lim_{n \to \infty} B_n = B \end{cases}$$

and if the limit $\lim_{n \to \infty} a_{n+1}/a_n = \lambda$ exists, then λ must satisfy (divide (B.1) by a_{n-1} and take the limit $n \to \infty$)

$$\frac{a_{n+1}}{a_n} \frac{a_n}{a_{n-1}} = \frac{a_{n+1}}{a_{n-1}} = A_n \frac{a_n}{a_{n-1}} + B_n \quad \Longrightarrow \quad \lambda^2 = A\lambda + B$$

The following theorem is instrumental for the existence of the limit (proof omitted, see also [6, Sec. 8.5]):

Theorem B.1 (Poincaré, Perron). *Let* $\{x_n\}_{n=0}^{\infty}$ *be a sequence generated by the recursion relation*

$$\begin{cases} x_{n+1} = A_n x_n + B_n x_{n-1}, & n \in \mathbb{Z}_+ \\ x_1 = A_0 x_0 \end{cases}$$

where the coefficients have well-defined limits as $n \to \infty$, *i.e.,*

$$\begin{cases} \lim_{n \to \infty} A_n = A \\ \lim_{n \to \infty} B_n = B \end{cases}$$

and, moreover, that all $B_n \neq 0$ *for all* $n \in \mathbb{Z}_+$. *The roots to the characteristic equation*

$$\lambda^2 = A\lambda + B$$

are denoted $\lambda = \lambda_1, \lambda_2$, *i.e.,*

$$\lambda_{1,2} = \frac{A \pm (A^2 + 4B)^{1/2}}{2}$$

---

[1] Henri Poincaré (1854–1912), French mathematician and theoretical physicist.
[2] Oskar Perron (1880–1975), German mathematician.

We assume these roots have different moduli, say $|\lambda_1| < |\lambda_2|$.

Then there exists a fundamental set of solutions,[3] $\{a_n\}_{n=0}^{\infty}$ and $\{b_n\}_{n=0}^{\infty}$, such that

$$\lim_{n \to \infty} \frac{a_{n+1}}{a_n} = \lambda_2, \quad and \quad \lim_{n \to \infty} \frac{b_{n+1}}{b_n} = \lambda_1$$

As a consequence of the Poincaré–Perron theorem, Theorem B.1, we have

Theorem B.2. *Let $\{x_n\}_{n=0}^{\infty}$ be a sequence generated by the recursion relation*

$$\begin{cases} x_{n+1} = A_n x_n + B_n x_{n-1}, & n \in \mathbb{Z}_+ \\ x_1 = A_0 x_0 \end{cases} \tag{B.2}$$

and

$$\begin{cases} \lim_{n \to \infty} A_n = A \\ \lim_{n \to \infty} B_n = B \end{cases}$$

Assume the roots, $\lambda = \lambda_1, \lambda_2$, of the characteristic equation

$$\lambda^2 = A\lambda + B \tag{B.3}$$

i.e.,

$$\lambda_{1,2} = \frac{A \pm (A^2 + 4B)^{1/2}}{2}$$

have different moduli, say $|\lambda_1| < |\lambda_2|$.

Then the limit $\lim_{n \to \infty} x_{n+1}/x_n$ always exists, and it is

$$\lim_{n \to \infty} \frac{x_{n+1}}{x_n} = \lambda_2$$

except when the solution is a multiple of $\{b_n\}_{n=0}^{\infty}$, given in Theorem B.1. Then the limit is

$$\lim_{n \to \infty} \frac{x_{n+1}}{x_n} = \lambda_1$$

Proof. As a consequence of Theorem B.1, there exists a fundamental set of solutions, $\{a_n\}_{n=0}^{\infty}$ and $\{b_n\}_{n=0}^{\infty}$, such that

$$\lim_{n \to \infty} \frac{a_{n+1}}{a_n} = \lambda_2, \quad and \quad \lim_{n \to \infty} \frac{b_{n+1}}{b_n} = \lambda_1$$

Let μ_1 and μ_2 be real numbers, such that $|\lambda_1| < \mu_1 < \mu_2 < |\lambda_2|$. Then there exists an integer N, such that

$$\left| \frac{a_{n+1}}{a_n} \right| \geq \mu_2, \quad and \quad \left| \frac{b_{n+1}}{b_n} \right| \leq \mu_1, \quad n \geq N$$

---

[3] The existence of a fundamental set implies that every solution of the recursion relation can be found as a linear combination of the elements in this set.

which imply

$$|a_n| = \left|\frac{a_n}{a_{n-1}}\right| \left|\frac{a_{n-1}}{a_{n-2}}\right| \cdots \left|\frac{a_{N+1}}{a_N}\right| |a_N| \geq \mu_2^{n-N} |a_N|, \quad n \geq N$$

and

$$|b_n| = \left|\frac{b_n}{b_{n-1}}\right| \left|\frac{b_{n-1}}{b_{n-2}}\right| \cdots \left|\frac{b_{N+1}}{b_N}\right| |b_N| \leq \mu_1^{n-N} |b_N|, \quad n \geq N$$

The quotient b_n/a_n then converges to zero, i.e.,

$$\lim_{n \to \infty} \left|\frac{b_n}{a_n}\right| \leq \lim_{n \to \infty} \left(\frac{\mu_1}{\mu_2}\right)^{n-N} \left|\frac{b_N}{a_N}\right| = 0$$

We say that the solution $\{b_n\}_{n=0}^{\infty}$ is a minimal solution (sequence). We also have

$$\lim_{n \to \infty} \left|\frac{b_{n+1}}{a_n}\right| \leq \lim_{n \to \infty} \left(\frac{\mu_1}{\mu_2}\right)^{n-N} \mu_1 \left|\frac{b_N}{a_N}\right| = 0$$

The general solution to (B.2) is $x_n = \alpha a_n + \beta b_n$ for some constants α and β. Then

$$\lim_{n \to \infty} \frac{x_{n+1}}{x_n} = \lim_{n \to \infty} \frac{\alpha a_{n+1} + \beta b_{n+1}}{\alpha a_n + \beta b_n} = \lim_{n \to \infty} \frac{\alpha a_{n+1}/a_n + \beta b_{n+1}/a_n}{\alpha + \beta b_n/a_n} = \lambda_2$$

This is the general limit value except when $\alpha = 0$ and $x_n = \beta b_n$. Then the limit is

$$\lim_{n \to \infty} \frac{x_{n+1}}{x_n} = \lambda_1$$

and the theorem is proved.   □

The proof of the following theorem is presented on page 402 in Ref. 6, and it provides conditions on the coefficients A_n and B_n for the solution to be minimal.

Theorem B.3 (Pincherle[4]). *Let $\{x_n\}_{n=0}^{\infty}$ be a sequence generated by the recursion relation*

$$\begin{cases} x_{n+1} = A_n x_n + B_n x_{n-1}, & n \in \mathbb{Z}_+ \\ x_1 = A_0 x_0 \end{cases}$$

where the coefficients have well-defined limits as $n \to \infty$, i.e.,

$$\begin{cases} \lim_{n \to \infty} A_n = A \\ \lim_{n \to \infty} B_n = B \end{cases}$$

Then the continued fraction

---

[4] Salvatore Pincherle (1853–1936), Italian mathematician.

$$A_0 + \cfrac{B_1}{A_1 + \cfrac{B_2}{A_2 + \cfrac{B_3}{A_3 + \dots}}}$$

or in a compact notation

$$A_0 + \overset{\infty}{\underset{n=1}{\Phi}} \frac{B_n}{A_n}$$

converges if and only if the sequence $\{x_n\}_{n=0}^{\infty}$ $(x_0 \neq 0)$ *is a minimal solution (sequence), i.e.,*

$$\lim_{n \to \infty} \frac{x_n}{a_n} = 0$$

where the sequence a_n *is given in Theorem B.1. Moreover, in the case of convergence*

$$\cfrac{B_1}{A_1 + \cfrac{B_2}{A_2 + \cfrac{B_3}{A_3 + \dots}}} = -\frac{x_1}{x_0} = -A_0 \tag{B.4}$$

B.3 Asymptotic behavior of recursion relations

The asymptotic behavior of the sequence generated by a special type of recursion relation, (B.1), for large values of n is addressed in this section.[5] The results are not so general as the results in Section B.2, but they suffice for our needs, the analysis is self-contained, and it uses only standard analysis arguments.

The start of the analysis is motivated by the following simple example.

Example B.1. If $A_n = 1 + \alpha$ and $B_n = -\alpha$, $n \in \mathbb{Z}_+$, where α is independent of n, the sequences $a_n = a_0$ $(A_0 = 1)$ and $a_n = \alpha^n a_0$ $(A_0 = \alpha)$ are solutions to the recursion relation (B.1). If $|\alpha| < 1$, the solutions converge to a_0 and 0, respectively. ■

For $a_n = a_0$ to be a solution to the recursion relation (B.1), it suffices to require $A_n + B_n = 1$, $n \in \mathbb{Z}_+$ and $A_0 = 1$. With this observation in mind, we anticipate that the size of $A_n + B_n - 1$ as $n \to \infty$ is essential for the convergence. That this really is the case is shown in this section.

We prefer to collect the results in two lemmas, a theorem, and a corollary. We start by proving the lemmas.

Lemma B.3. *Let* $\{a_n\}_{n=0}^{\infty}$ *be a sequence generated by the recursion relation*

$$\begin{cases} a_{n+1} = A_n a_n + B_n a_{n-1}, & n \in \mathbb{Z}_+ \\ a_1 = A_0 a_0 \end{cases}$$

---

[5] The idea behind the approach presented here is due to Anders Melin.

where the coefficients A_n and B_n satisfy

$$\begin{cases} \lim_{n\to\infty} A_n = 1+\alpha \\ \lim_{n\to\infty} B_n = -\alpha \end{cases}$$

and $|\alpha| < 1$. Moreover, let $\lambda > 1$ be a real number, such that

$$R_n = A_n + B_n - 1 = O(n^{-\lambda})$$

Then, for all initial values a_0, the sequence $\{a_n\}_{n=0}^{\infty}$ converges to a limit d, and

$$a_n = d + O(n^{-\lambda+1}), \quad n \to \infty$$

Proof. Rewrite the recursion relation as

$$a_{n+1} - a_n = R_n a_n - B_n(a_n - a_{n-1}), \qquad n \in \mathbb{Z}_+ \qquad (B.5)$$

and we start by proving that the sequence $\{a_n\}_{n=0}^{\infty}$ is bounded.

Let N be a positive integer such that

$$|B_n| \leq c = \frac{1+|\alpha|}{2}, \quad n \geq N$$

Notice that $0 < c < 1$ with the assumptions made in the lemma. We also have

$$|R_n| \leq \frac{C}{n^{\lambda}}, \quad n \geq N$$

for some constant C. Define

$$\varepsilon_n = |a_{n+1} - a_n|, \qquad n \in \mathbb{N}$$

Then from (B.5)

$$\varepsilon_n \leq \frac{C}{n^{\lambda}}|a_n| + c\varepsilon_{n-1}, \quad n \geq N \qquad (B.6)$$

and

$$\sum_{k=N}^{n} \varepsilon_k \leq \sum_{k=N}^{n} \frac{C}{k^{\lambda}}|a_k| + c\sum_{k=N}^{n} \varepsilon_k + c\varepsilon_{N-1}, \quad n \geq N$$

From this expression we conclude that

$$\sum_{k=N}^{n} \varepsilon_k \leq C_1 \sum_{k=N}^{n} k^{-\lambda}|a_k| + C_2, \quad n \geq N \qquad (B.7)$$

where

$$\begin{cases} C_1 = \dfrac{C}{1-c} > 0 \\ C_2 = \dfrac{c\mathcal{E}_{N-1}}{1-c} > 0 \end{cases}$$

Introduce the notation

$$\hat{a}_n = \max_{0 \le k \le n} |a_k|, \qquad n \in \mathbb{N}$$

and use the inequality

$$\hat{a}_{n+1} = \max_{0 \le k \le n} \{|a_k|, \underbrace{|a_{n+1}|}_{\le \mathcal{E}_n + a_n}\} \le \mathcal{E}_n + \hat{a}_n$$

to obtain[6]

$$\hat{a}_{n+1} \le \hat{a}_N + \sum_{k=0}^{n-N} \mathcal{E}_{n-k} \le \hat{a}_N + \sum_{k=N}^{n} \mathcal{E}_k, \quad n \ge N$$

Use (B.7), and we obtain

$$\hat{a}_{n+1} \le C_1 \sum_{k=N}^{n} k^{-\lambda} |a_k| + C_2 + \hat{a}_N \le \hat{a}_n C_1 \sum_{k=N}^{n} k^{-\lambda} + C_3, \quad n \ge N$$

where $C_3 = C_2 + \hat{a}_N$. Choose the integer N large enough so that

$$C_1 \sum_{k=N}^{\infty} k^{-\lambda} \le \frac{1}{2}$$

and we get

$$\hat{a}_{n+1} \le \frac{1}{2} \hat{a}_n + C_3, \quad n \ge N$$

By the use of the result in footnote 6, we get

$$\hat{a}_{n+1} \le \frac{1}{2^{n-N+1}} \hat{a}_N + C_3 \sum_{k=0}^{n-N} \frac{1}{2^k}, \quad n \ge N$$

---

[6] In this section we make frequent use of inequalities of the type

$$x_{n+1} \le a x_n + b_n, \quad n \ge N$$

By induction over m, we get

$$x_{n+1} \le a^{m+1} x_{n-m} + \sum_{k=0}^{m} a^k b_{n-k}, \quad 0 \le m \le n - N$$

$$x_{n+1} \le a^{n-N+1} x_N + \sum_{k=0}^{n-N} a^k b_{n-k}, \quad n \ge N$$

or

$$\hat{a}_{n+1} \le \hat{a}_N + 2C_3, \quad n \ge N$$

which proves that the sequence $\{a_n\}_{n=0}^{\infty}$ is bounded, and the first part of the proof is completed.

We are now able to use the boundedness of the sequence $\{a_n\}_{n=0}^{\infty}$ to prove that this sequence is a Cauchy sequence of complex numbers, and therefore has a limit. Use (B.6), and we get for a suitable constant C

$$\varepsilon_n \le c\varepsilon_{n-1} + \frac{C}{n^\lambda}, \quad n \ge N$$

Use the result in footnote 6 to get

$$\varepsilon_n \le c^{n-N+1}\varepsilon_{N-1} + C\sum_{k=N}^{n}\frac{c^{n-k}}{k^\lambda}, \quad n \ge N \tag{B.8}$$

However, the sequence

$$S_n = \sum_{k=1}^{n}\frac{c^n n^\lambda}{c^k k^\lambda}$$

is bounded, see Lemma B.8 on page 197, which implies that

$$\sum_{k=1}^{n}\frac{c^{n-k}}{k^\lambda} \le \frac{C'}{n^\lambda}$$

for a suitable constant C'. Therefore, ε_n in (B.8) is estimated as

$$\varepsilon_n \le c^{n-N+1}\varepsilon_{N-1} + \frac{C''}{n^\lambda}, \quad n \ge N$$

which can be arbitrary small for all $n \ge N'$, provided $N' > N$ is large enough.

Moreover, the sequence $\{a_n\}_{n=0}^{\infty}$ is a Cauchy sequence, since we have from the results above that[7]

$$|a_n - a_m| \le \sum_{k=m}^{n-1}\varepsilon_k \le \frac{c^{m-N+1} - c^{n-N+1}}{1-c}\varepsilon_{N-1} + C''\sum_{k=m}^{n-1}\frac{1}{k^\lambda}$$

$$\le \frac{c^{m-N+1} - c^{n-N+1}}{1-c}\varepsilon_{N-1} + \frac{\lambda}{\lambda-1}\frac{C''}{m^{\lambda-1}}, \quad n > m \ge N' > N$$

---

[7] For example, use the estimate (B.9) ($n > m \ge 1$ and $\lambda > 1$)

$$\sum_{k=m}^{n-1}\frac{1}{k^\lambda} \le \frac{1}{m^\lambda} + \int_m^{n-1}\frac{dx}{x^\lambda} = \frac{1}{m^\lambda} + \frac{1}{\lambda-1}\left(\frac{1}{m^{\lambda-1}} - \frac{1}{(n-1)^{\lambda-1}}\right)$$

$$\le \left(1 + \frac{1}{\lambda-1}\right)\frac{1}{m^{\lambda-1}} \le \frac{\lambda}{\lambda-1}\frac{1}{m^{\lambda-1}}$$

which can be made arbitrarily small for $n > m \geq N'$, provided N' is chosen sufficiently large. If the limit of the sequence $\{a_n\}_{n=0}^{\infty}$ is denoted d, we also have from above that

$$|a_n - d| \leq \sum_{k=n}^{\infty} \varepsilon_k \leq \frac{c^{n-N+1}}{1-c} \varepsilon_{N-1} + \frac{\lambda}{\lambda-1} \frac{c''}{n^{\lambda-1}}, \quad n \geq N'$$

Therefore, $a_n = d + O(n^{-\lambda+1})$, and the lemma is proved. $\quad\square$

The next lemma shows how the convergence of specific combination of the coefficients in the recursion relation can be improved.

Lemma B.4. *If $\{a_n\}_{n=0}^{\infty}$ is a sequence generated by the recursion relation*

$$\begin{cases} a_{n+1} = A_n a_n + B_n a_{n-1}, & n \in \mathbb{Z}_+ \\ a_1 = A_0 a_0 \end{cases}$$

where the coefficients A_n and B_n satisfy

$$\begin{cases} A_n = \beta_0 + \dfrac{\beta_1}{n} + \dfrac{\beta_2}{n^2} + O(1/n^3) \\ B_n = \gamma_0 + \dfrac{\gamma_1}{n} + \dfrac{\gamma_2}{n^2} + O(1/n^3) \end{cases} \quad \text{as } n \to \infty$$

where it is assumed that $\beta_0 + 2\gamma_0 \neq 0$. Then the sequence $\{a_n'\}_{n=0}^{\infty}$ defined by

$$a_n' = a_n \prod_{k=1}^{n} \left(1 + \frac{c_1}{k} + \frac{c_2}{k^2}\right), \quad n \in \mathbb{Z}_+, \quad a_0' = a_0$$

where

$$c_1 = -\frac{\beta_1 + \gamma_1}{\beta_0 + 2\gamma_0}, \quad c_2 = \frac{c_1(\beta_0 - 2\gamma_1 - \beta_1 + \gamma_0 - c_1\gamma_0) - \gamma_2 - \beta_2}{\beta_0 + 2\gamma_0}$$

satisfies the recursion relation

$$\begin{cases} a_{n+1}' = A_n' a_n' + B_n' a_{n-1}', & n \in \mathbb{Z}_+ \\ a_1' = A_0' a_0' \end{cases}$$

where $A_0' = A_0$ and

$$\begin{cases} A_n' = A_n \left(1 + \dfrac{c_1}{n+1} + \dfrac{c_2}{(n+1)^2}\right) \\ B_n' = B_n \left(1 + \dfrac{c_1}{n+1} + \dfrac{c_2}{(n+1)^2}\right) \left(1 + \dfrac{c_1}{n} + \dfrac{c_2}{n^2}\right) \end{cases} \quad n \in \mathbb{Z}_+$$

Moreover, the coefficients A_n' and B_n' satisfy

$$A'_n + B'_n - \beta_0 - \gamma_0 = O(n^{-3})$$

Proof. The idea behind the proof is to replace the sequence $\{a_n\}_{n=0}^{\infty}$ by the sequence $\{a'_n\}_{n=0}^{\infty}$ defined by

$$a'_n = a_n \prod_{k=1}^{n}\left(1 + \frac{c_1}{k} + \frac{c_2}{k^2}\right), \quad n \in \mathbb{Z}_+, \quad a'_0 = a_0$$

where c_1 and c_2 are to be determined, such that the claims of the lemma are satisfied.
The new sequence $\{a'_n\}_{n=0}^{\infty}$ satisfies

$$\begin{cases} a'_{n+1} = A'_n a'_n + B'_n a'_{n-1}, & n \in \mathbb{Z}_+ \\ a'_1 = A'_0 a'_0 \end{cases}$$

where $A'_0 = A_0$ and

$$\begin{cases} A'_n = A_n \left(1 + \dfrac{c_1}{n+1} + \dfrac{c_2}{(n+1)^2}\right) \\[2mm] B'_n = B_n \left(1 + \dfrac{c_1}{n+1} + \dfrac{c_2}{(n+1)^2}\right)\left(1 + \dfrac{c_1}{n} + \dfrac{c_2}{n^2}\right) \end{cases} \quad n \in \mathbb{Z}_+$$

If we insert the expansions of the coefficients A_n and B_n, we get

$$\begin{cases} A'_n = \left(\beta_0 + \dfrac{\beta_1}{n} + \dfrac{\beta_2}{n^2} + O(1/n^3)\right)\left(1 + \dfrac{c_1}{n+1} + \dfrac{c_2}{(n+1)^2}\right) \\[2mm] \quad = \beta_0 + \dfrac{\beta_1 + \beta_0 c_1}{n} + \dfrac{\beta_1 c_1 + \beta_2 + \beta_0 c_2 - \beta_0 c_1}{n^2} + O(1/n^3) \\[3mm] B'_n = \left(\gamma_0 + \dfrac{\gamma_1}{n} + \dfrac{\gamma_2}{n^2} + O(1/n^3)\right)\left(1 + \dfrac{c_1}{n+1} + \dfrac{c_2}{(n+1)^2}\right)\left(1 + \dfrac{c_1}{n} + \dfrac{c_2}{n^2}\right) \\[2mm] \quad = \gamma_0 + \dfrac{\gamma_1 + 2\gamma_0 c_1}{n} + \dfrac{\gamma_2 + \gamma_0(2c_2 + c_1^2 - c_1) + 2\gamma_1 c_1}{n^2} + O(1/n^3) \end{cases}$$

We also denote

$$h_n = A'_n + B'_n - \beta_0 - \gamma_0$$

and our aim is to choose c_1 and c_2 such that $h_n = O(n^{-3})$.

$$h_n = \frac{\beta_1 + \beta_0 c_1 + \gamma_1 + 2\gamma_0 c_1}{n}$$
$$+ \frac{\beta_1 c_1 + \beta_2 + \beta_0 c_2 - \beta_0 c_1 + \gamma_2 + \gamma_0(2c_2 + c_1^2 - c_1) + 2\gamma_1 c_1}{n^2} + O(1/n^3)$$

and we see that the conditions in the lemma are proved, if we choose c_1 and c_2 as

$$\beta_1 + c_1(\beta_0 + 2\gamma_0) + \gamma_1 = 0 \quad \Longrightarrow \quad c_1 = -\frac{\beta_1 + \gamma_1}{\beta_0 + 2\gamma_0}$$

and
$$\beta_1 c_1 + \beta_2 + \beta_0 c_2 - \beta_0 c_1 + \gamma_2 + \gamma_0(2c_2 + c_1^2 - c_1) + 2\gamma_1 c_1 = 0$$

which we simplify to

$$c_2 = \frac{c_1(\beta_0 - 2\gamma_1 - \beta_1 + \gamma_0 - c_1\gamma_0) - \gamma_2 - \beta_2}{\beta_0 + 2\gamma_0}$$

☐

The main theorem of the section that proves convergence of sequences can now be formulated.

Theorem B.4. *Let $\{a_n\}_{n=0}^\infty$ be a sequence generated by the recursion relation*

$$\begin{cases} a_{n+1} = A_n a_n + B_n a_{n-1}, & n \in \mathbb{Z}_+ \\ a_1 = A_0 a_0 \end{cases}$$

where the coefficients A_n and B_n satisfy

$$\begin{cases} A_n = 1 + \alpha + \dfrac{\beta_1}{n} + \dfrac{\beta_2}{n^2} + O(1/n^3) \\ B_n = -\alpha + \dfrac{\gamma_1}{n} + \dfrac{\gamma_2}{n^2} + O(1/n^3) \end{cases} \quad as \ n \to \infty$$

and $|\alpha| < 1$. Then, provided the sequence $\{a_n\}_{n=0}^\infty$ does not converge to zero, the sequence for large n behaves as

$$a_n = C n^{(\beta_1 + \gamma_1)/(1-\alpha)} \left(1 + \frac{c}{n} + O(1/n^2)\right), \quad n \to \infty$$

for some constants C and c.

Proof. We prove the theorem by applying Lemmas B.3 and B.4. With the notation and the results of these lemmas, there exists a sequence $\{a_n'\}_{n=0}^\infty$ defined by (notice that $1 + \alpha + 2(-\alpha) = 1 - \alpha \neq 0$)

$$a_n' = a_n \prod_{k=1}^n \left(1 + \frac{c_1}{k} + \frac{c_2}{k^2}\right), \qquad a_0' = a_0$$

$$c_1 = -\frac{\beta_1 + \gamma_1}{1 - \alpha}, \qquad c_2 = \frac{c_1(1 - 2\gamma_1 - \beta_1 + c_1\alpha) - \gamma_2 - \beta_2}{1 - \alpha}$$

which is converging to a limit $d \neq 0$, such that ($\lambda = 3$)

$$a_n' - d = O(n^{-2}), \quad as \ n \to \infty$$

Note that required asymptotic behavior needed in Lemma B.4 is one order higher, here $\lambda = 3$, than the result of Lemma B.3.

We write the original sequence as

$$a_n = a'_n e^{q_n} = d e^{q_n} \left(1 + O(n^{-2})\right)$$

where

$$q_n = -\sum_{k=1}^{n} \ln\left(1 + \frac{c_1}{k} + \frac{c_2}{k^2}\right)$$

We proceed by finding an asymptotic expansion of q_n valid for large values of n. Start with the Taylor expansion

$$\ln\left(1 + c_1 x + c_2 x^2\right) = F(x) = \sum_{m=1}^{\infty} \frac{F^{(m)}(0)}{m!} x^m, \quad 0 \le x \le 1$$

where $F'(0) = c_1$. We also need the asymptotic expansions of the following sums:

$$\sum_{k=1}^{n} \frac{1}{k} = \ln n + c_{1,0} + \frac{c_{1,1}}{n} + O(n^{-2})$$

and

$$\sum_{k=1}^{n} \frac{1}{k^m} = c_{m,0} + \frac{c_{m,m-1}}{n^{m-1}} + O(n^{-m}) = c_{m,0} + \frac{c_{m,m-1}}{n^{m-1}} + O(n^{-2}), \quad m \ge 2$$

We also define $c_{m,j} = 0$ for $j = 1, 2, \ldots, m-2$, $m \ge 3$. Specific values of the constants are, see Lemmas B.6 and B.7

$$\begin{cases} c_{1,0} = \gamma \\ c_{1,1} = \dfrac{1}{2} \end{cases} \qquad \begin{cases} c_{m,0} = \zeta(m) \\ c_{m,m-1} = \dfrac{1}{1-m} \end{cases} \quad m \ge 2$$

where γ is the Euler–Mascheroni constant defined in (A.18) on page 176, and $\zeta(m) = \sum_{n=1}^{\infty} n^{-m}$ is the Riemann zeta function. We get

$$\sum_{k=1}^{n} \ln\left(1 + \frac{c_1}{k} + \frac{c_2}{k^2}\right) = \sum_{k=1}^{n} \sum_{m=1}^{\infty} \frac{F^{(m)}(0)}{m!} k^{-m} = \sum_{m=1}^{\infty} \frac{F^{(m)}(0)}{m!} \sum_{k=1}^{n} k^{-m}$$

$$= F'(0) \ln n + \sum_{m=1}^{\infty} \frac{F^{(m)}(0)}{m!} \left(c_{m,0} + \frac{c_{m,1}}{n} + O(n^{-2})\right)$$

$$= c_1 \ln n + C_0 + \frac{C_1}{n} + O(n^{-2})$$

where $O(n^{-2})$ has the meaning $\left|O(n^{-2})\right| \le C n^{-2}$, with C independent of m. Moreover, the constants C_0 and C_1 are

$$\begin{cases} C_0 = \sum_{m=1}^{\infty} \frac{F^{(m)}(0)}{m!} c_{m,0} \\[2ex] C_1 = \sum_{m=1}^{\infty} \frac{F^{(m)}(0)}{m!} c_{m,1} = \sum_{m=1}^{2} \frac{F^{(m)}(0)}{m!} c_{m,1} \end{cases}$$

and consequently

$$q_n = -c_1 \ln n - C_0 - \frac{C_1}{n} + O(n^{-2})$$

The sequence of interest then is

$$a_n = d e^{-c_1 \ln n - C_0 - \frac{C_1}{n} + O(n^{-2})} \left(1 + O(n^{-2})\right) = C n^{-c_1} \left(1 + \frac{c}{n} + O(n^{-2})\right)$$

for some constants C and c, which concludes the proof. □

Corollary B.1. *With the assumptions in Theorem B.4, we have that*

$$\frac{a_{n+1}}{a_n} = 1 + \frac{\beta_1 + \gamma_1}{n(1-\alpha)} + O(1/n^2), \quad \text{as } n \to \infty$$

Proof. Use the result of Theorem B.4, i.e.,

$$a_n = C n^{(\beta_1 + \gamma_1)/(1-\alpha)} \left(1 + \frac{c}{n} + O(1/n^2)\right), \quad n \to \infty$$

which implies that

$$\frac{a_{n+1}}{a_n} = \left(\frac{n+1}{n}\right)^{(\beta_1+\gamma_1)/(1-\alpha)} \frac{1 + \frac{c}{n+1} + O(1/n^2)}{1 + \frac{c}{n} + O(1/n^2)} = 1 + \frac{\beta_1 + \gamma_1}{n(1-\alpha)} + O(1/n^2)$$

and the corollary follows. □

Notice that the result of Corollary B.1 is consistent with the result of Theorem B.1 since $\lambda_{1,2} = 1, \alpha$.

B.4 Estimates of some sequences and series

For convenience we here collect a series of lemmas on estimates of sequences and series that are used above. We start by stating a general lemma that relates a series to the corresponding integral.

Lemma B.5. *Let the real-valued function $f(x)$ be non-increasing in the interval $[m,n]$, $m,n \in \mathbb{N}$. Then*

$$f(n) \le \sum_{k=m}^{n} f(k) - \int_{m}^{n} f(x)\,dx \le f(m), \quad n \ge m \tag{B.9}$$

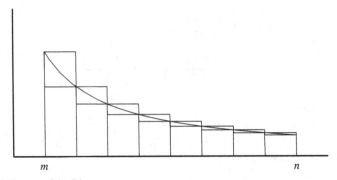

Fig. B.1 Estimates of the Riemann sum.

Proof. This lemma is easily proved by estimating the Riemann sum of the integral, see Figure B.1. We have for integers $m < n$

$$\sum_{k=m+1}^{n} f(k) \leq \int_{m}^{n} f(x)\, dx \leq \sum_{k=m}^{n-1} f(k)$$

which is identical to the statement in the lemma. □

B.4.1 Riemann zeta function

Lemma B.6. *Let* $\lambda > 1$ *be a real number. Then, the partial sum*

$$s_n = \sum_{k=1}^{n} \frac{1}{k^{\lambda}}, \qquad n \in \mathbb{Z}_+$$

has the following asymptotic expansion:

$$s_n = s + \frac{n^{1-\lambda}}{1-\lambda} + O(n^{-\lambda})$$

where $s = \zeta(\lambda)$ *is the Riemann zeta function.*

Proof. The sequence s_n is increasing, since $s_{n+1} - s_n = (n+1)^{-\lambda} \geq 0$. Moreover, the sequence s_n is bounded from above, which is proved using (B.9) with $f(x) = x^{-\lambda}$ and $m = 1$.

$$0 \leq s_n \leq \int_{1}^{n} x^{-\lambda}\, dx + 1 = \frac{1}{\lambda-1}\left(1 - n^{-\lambda+1}\right) + 1 < \frac{1}{\lambda-1} + 1 = \frac{\lambda}{\lambda-1}$$

The sequence s_n therefore has a limit s.

The remaining part of the lemma is proved using the notation

$$q_n = n^{\lambda-1}(s - s_n) = n^{\lambda-1}\sum_{k=n+1}^{\infty}\frac{1}{k^{\lambda}} \geq 0$$

Use (B.9) with $f(x) = n^{\lambda-1}x^{-\lambda}$, $m = n+1$, and the upper limit approaching infinity. We get as $f(x) \to 0$ as $x \to \infty$,

$$0 \leq q_n - n^{\lambda-1}\int_{n+1}^{\infty} x^{-\lambda}\,dx \leq n^{\lambda-1}(n+1)^{-\lambda}$$

or by evaluating the integral

$$0 \leq q_n - \frac{1}{\lambda-1}\left(1+\frac{1}{n}\right)^{1-\lambda} \leq \frac{1}{n}\left(1+\frac{1}{n}\right)^{-\lambda}$$

which proves that

$$q_n = \frac{1}{\lambda-1} + O(n^{-1}), \quad \text{as } n \to \infty$$

and

$$s_n = s - q_n n^{1-\lambda} = s + \frac{n^{1-\lambda}}{1-\lambda} + O(n^{-\lambda})$$

and the lemma is proved.   □

B.4.2 The sum $\sum_{k=1}^{n} k^{-1}$

The following lemma shows the asymptotic behavior of the series $\sum_{k=1}^{n} k^{-1}$, which also is used above:

Lemma B.7. *Define the partial sums*

$$s_n = \sum_{k=1}^{n}\frac{1}{k}, \quad n \in \mathbb{Z}_+$$

Then the sequence s_n has the asymptotic expansion

$$s_n = \ln n + \gamma + \frac{1}{2n} + O(n^{-2})$$

where the Euler–Mascheroni constant γ is

$$\gamma = \lim_{n\to\infty}\left\{\sum_{k=1}^{n}\frac{1}{k} - \ln n\right\}$$

A more detailed treatment of the Euler–Mascheroni constant is presented in Appendix A on page 176.

Proof. Introduce the notation

$$q_n = \sum_{k=1}^{n} \frac{1}{k} - \int_1^n \frac{dx}{x} = s_n - \ln n$$

The sequence q_n is a decreasing sequence since

$$q_n - q_{n-1} = \frac{1}{n} - \int_{n-1}^n \frac{dx}{x} = \frac{1}{n} + \ln\left(1 - \frac{1}{n}\right) \le 0$$

since $\ln(1+x) \le x$ for $x \ge -1$. Utilize the inequality (B.9) with $f(x) = 1/x$ and $m = 1$,

$$0 < \frac{1}{n} \le q_n \le 1, \qquad n \in \mathbb{Z}_+ \tag{B.10}$$

which shows that the decreasing sequence q_n is bounded from below, and therefore the limit $q_n \to \gamma$ as $n \to \infty$ exists, i.e.,

$$\gamma = \lim_{n \to \infty} \left\{ \sum_{k=1}^{n} \frac{1}{k} - \ln n \right\} = \sum_{k=1}^{\infty} \left\{ \frac{1}{k} - \ln\left(1 + \frac{1}{k}\right) \right\}$$

where we in the last equality have used (A.19) on page 176.

To proceed, rewrite

$$\sum_{k=1}^{n} \left\{ \frac{1}{k} - \ln\left(1 + \frac{1}{k}\right) \right\} = \sum_{k=1}^{n} \frac{1}{k} - \ln\left(\prod_{k=1}^{n} \frac{k+1}{k} \right) = \sum_{k=1}^{n} \frac{1}{k} - \ln(n+1)$$

$$= \sum_{k=1}^{n} \frac{1}{k} - \ln n + \ln \frac{n}{n+1} = q_n - \ln\left(1 + \frac{1}{n}\right)$$

and we get

$$\gamma - q_n + \ln\left(1 + \frac{1}{n}\right) = \sum_{k=n+1}^{\infty} \left\{ \frac{1}{k} - \ln\left(1 + \frac{1}{k}\right) \right\}$$

Now use the inequality (B.9) with

$$f(x) = n\left(\frac{1}{x} - \ln\left(1 + \frac{1}{x}\right) \right) \ge 0, \quad x > 1$$

Letting the upper limit in (B.9) approach infinity, $m = n+1$, and using $f(x) \to 0$ as $x \to \infty$ for each n, we get

$$0 \le n\left\{ \gamma - q_n + \ln\left(1 + \frac{1}{n}\right) - \int_{n+1}^{\infty} \left\{ \frac{1}{x} - \ln\left(1 + \frac{1}{x}\right) \right\} dx \right\}$$

$$\le \frac{n}{n+1} - n\ln\left(1 + \frac{1}{n+1}\right) = O(1/n)$$

since $\ln(1+x) = x - x^2/2 + O(x^3)$ as $x \to 0$. We conclude that

$$n(\gamma - q_n) + 1 = n \int_{n+1}^{\infty} \left\{ \frac{1}{x} - \ln\left(1 + \frac{1}{x}\right) \right\} dx + O(n^{-1})$$

The integral can be evaluated as

$$n \int_{n+1}^{\infty} \left\{ \frac{1}{x} - \ln\left(1 + \frac{1}{x}\right) \right\} dx = n \int_0^{1/(n+1)} \frac{x - \ln(1+x)}{x^2} dx$$

$$= n \int_0^{1/(n+1)} \frac{1}{2} dx + O(n^{-1}) = \frac{1}{2} + O(n^{-1})$$

We finally get

$$q_n = s_n - \ln n = \gamma + \frac{1}{2n} + O(n^{-2})$$

and

$$s_n = \ln n + \gamma + \frac{1}{2n} + O(n^{-2})$$

and the lemma is proved. $\square$

B.4.3 Convergence of a sequence

Lemma B.8. *Let $c \in (0,1)$ and $\lambda \in \mathbb{C}$. Then the sequence*

$$s_n = \sum_{k=1}^{n} c^{n-k} \left(\frac{n}{k}\right)^{\lambda}, \quad n \in \mathbb{Z}_+$$

is convergent with limit $s = 1/(1-c)$.

Proof. The sequence $\{s_n\}_{n=1}^{\infty}$ satisfies

$$s_{n+1} = 1 + \frac{c(n+1)^{\lambda}}{n^{\lambda}} \sum_{k=1}^{n} c^{n-k} \left(\frac{n}{k}\right)^{\lambda} = 1 + c \left(1 + \frac{1}{n}\right)^{\lambda} s_n = 1 + b_n s_n, \quad n \in \mathbb{Z}_+$$

where

$$b_n = c \left(1 + \frac{1}{n}\right)^{\lambda} \to c, \quad n \to \infty$$

If the sequence $\{s_n\}_{n=1}^{\infty}$ has a limit s, then it must satisfy

$$s = 1 + cs \quad \Rightarrow \quad s = \frac{1}{1-c} > 1$$

Therefore, write the sequence $\{s_n\}_{n=1}^{\infty}$ as

$$s_n = s + t_n, \quad n \in \mathbb{Z}_+$$

and the lemma follows, if we can prove that the sequence $\{t_n\}_{n=1}^{\infty}$ converges to zero.
The sequence $\{t_n\}_{n=1}^{\infty}$ satisfies

$$t_{n+1} = \frac{b_n - c}{1 - c} + b_n t_n = c \frac{\left(1 + \frac{1}{n}\right)^{\lambda} - 1}{1 - c} + b_n t_n, \quad n \in \mathbb{Z}_+$$

which implies that

$$|t_{n+1}| \leq \frac{C}{n} + |b_n| |t_n|, \quad n \in \mathbb{Z}_+$$

för some constant $C > 0$ since

$$\lim_{x \to 0} \frac{\left((1+x)^{\lambda} - 1\right)}{x} = \lambda$$

To proceed, introduce the constant $d = (1+c)/2 \in (c, 1)$, and let N be an integer
such that

$$|b_n| < d, \quad n \geq N$$

which is possible since $b_n \to c$ as $n \to \infty$. Then

$$|t_{n+1}| \leq \frac{C}{n} + d |t_n|, \quad n \geq N \tag{B.11}$$

Moreover, introduce

$$r_n = |t_n| - \frac{A}{n}, \quad n \geq N$$

where the positive constant A is chosen as

$$A = \frac{C(N+1)}{N(1-d) - d}$$

which is positive if $N > d/(1-d)$. Then from (B.11)

$$r_{n+1} \leq \frac{C}{n} - \frac{A}{n+1} + \frac{Ad}{n} + d r_n, \quad n \geq N$$

The chosen value of A implies that

$$\frac{C}{n} - \frac{A}{n+1} + \frac{Ad}{n} = \frac{C + dA - A\frac{n}{n+1}}{n} \leq \frac{C + dA - A\frac{N}{N+1}}{n} = 0, \quad n \geq N$$

and we obtain

$$r_{n+1} \leq d r_n, \quad n \geq N$$

Iteration of this expression gives

$$r_{n+1} \leq d^{n-N+1} r_N, \quad n \geq N$$

which proves that $r_n \to 0$, and consequently $|t_n| \to 0$, as $n \to \infty$, and the lemma is proved. $\square$

Appendix C
Partial fractions

A **rational function**, $r(z)$, is a quotient between two polynomials, $p(z)$ and $q(z)$,

$$r(z) = \frac{p(z)}{q(z)}$$

The rational function is defined in a domain in the complex z-plane, if we exclude the isolated zeros of $q(z)$, which we denote by

$$z_i, \qquad i = 1,2,3,\ldots,k$$

By polynomial division, it is always possible to write $r(z)$ as

$$r(z) = s_0(z) + \frac{p_0(z)}{q(z)}$$

where $s_0(z)$ is a polynomial, and where the degree of the polynomial $p_0(z)$ is less than the degree of the polynomial $q(z)$. The rational function $p_0(z)/q(z)$ is analytic at infinity.

If m_i is the multiplicity of z_i, then $q(z) = (z - z_i)^{m_i} q_i(z)$, where $q_i(z)$ is a polynomial and $q_i(z_i) \neq 0$. Let the principal parts of $r(z)$ be

$$s_i(z) = \sum_{j=1}^{m_i} \frac{a_{i,j}}{(z - z_i)^j}, \qquad i = 1,2,3,\ldots,k$$

The **partial fraction decomposition** of $r(z)$ then is

$$r(z) = \sum_{i=0}^{k} s_i(z)$$

Appendix D
Circles and ellipses in the complex plane

The circle and the ellipse occur frequently in the analysis in this book. In particular, we use circles and ellipses in the treatment of the convergence properties of the series solution of Heun's equation in Chapter 8. Their equations in the complex plane are reviewed in this appendix.

D.1 Equation of the circle

The equation for the circle, centered at $z = z_0$ and radius r, in the complex z-plane is

$$|z - z_0| = r$$

or equivalently

$$(z - z_0)(z^* - z_0^*) = r^2$$

where the star, $*$, denotes complex conjugate of the complex number. Denote $a = -z_0$ and $b = |z_0|^2 - r^2 \in \mathbb{R}$. The equation of the circle then is

$$zz^* + a^*z + az^* + b = 0, \quad b \text{ real}$$

Every circle in the complex plane has this form.

The analysis above implies that the equation

$$zz^* + \alpha z + \beta z^* + \gamma = 0 \tag{D.1}$$

represents a circle if and only if

$$\alpha = \beta^* \quad \text{and} \quad \alpha\beta - \gamma \text{ non-negative real number} \tag{D.2}$$

If these conditions are fulfilled, the circle has its center at $z = -\beta$ and the radius is $r = \sqrt{|\alpha|^2 - \gamma}$.

D.1.1 Harmonic circles

Let k be a real, positive number, and find the complex numbers z that satisfy

$$\left|\frac{z-a}{z-b}\right| = k \tag{D.3}$$

where a and b are distinct complex numbers, i.e., $a \neq b$. The problem has an equivalent formulation:

$$|z-a| = k|z-b|$$

and we observe that the problem is to find the complex numbers z, whose distances to the points a and b have a constant quotient k.

If $k = 1$, the solution is the straight line perpendicular to, and passing through the midpoint of the line connecting a and b.

We now show that the solution is a circle if $k \neq 1$. The equation is equivalent to

$$(z-a)(z^* - a^*) = k^2(z-b)(z^* - b^*)$$

or

$$(1-k^2)zz^* + (k^2 b^* - a^*)z + (k^2 b - a)z^* + |a|^2 - k^2|b|^2 = 0$$

This is the equation of a circle, see (D.1), provided the conditions in (D.2) are fulfilled. The first one in (D.2) is apparently satisfied, and the second is also satisfied, since

$$\frac{|k^2 b - a|^2}{(1-k^2)^2} - \frac{|a|^2 - k^2|b|^2}{1-k^2} = \frac{|k^2 b - a|^2 - (1-k^2)(|a|^2 - k^2|b|^2)}{(1-k^2)^2}$$
$$= \frac{k^2(|a|^2 + |b|^2 - 2\operatorname{Re} ab^*)}{(1-k^2)^2} = \frac{k^2|a-b|^2}{(1-k^2)^2} \geq 0$$

Equation (D.3) therefore is a circle centered at $z = z_0$ and radius r, where z_0 and r are given by

$$\begin{cases} z_0 = \dfrac{a - k^2 b}{1 - k^2} \\ r = \dfrac{k|a-b|}{|1-k^2|} \end{cases}$$

The center of the circle lies on the line connecting the points a and b, see Figure D.1, and the circle is called the harmonic circle to a and b. If $k < 1$, the circle encircles a, and if $z > 1$, the circle encircles the point b in the complex z-plane.

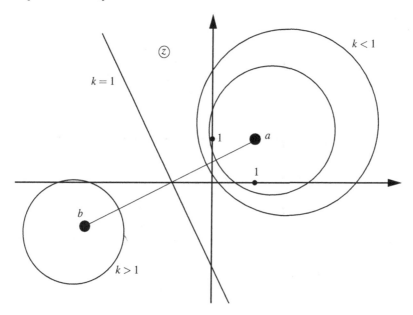

Fig. D.1 The harmonic circles in the complex z-plane. In the illustration $a = 1 + i$ and $b = -3 - i$. When $0 < k < 1$, the circles enclose the point a, and when $k > 1$, the circles enclose the point b.

D.2 Equation of the ellipse

The equation of the ellipse in the complex plane can take many forms. If the ellipse has foci at $z = z_1$ and $z = z_2$, and passes through $z = a$, one form of the equation is

$$|z - z_1| + |z - z_2| = |a - z_1| + |a - z_2|$$

This equation states that the sum of the distances from the point z in the complex z-plane to the foci is constant and is equal to the sum of the distances from the point a to the foci. In Theorem 8.2, we also use the following, less common, form of the ellipse:

Lemma D.1. *Let*
$$Z = \left(1 - z^{-1}\right)^{1/2}, \quad A = \left(1 - a^{-1}\right)^{1/2}$$

where the branches of the square roots are taken as the principal branch, $\mathrm{Re}\,Z > 0$ and $\mathrm{Re}\,A > 0$. Then, for a given a, the equation

$$\left|\frac{Z - 1}{Z + 1}\right| = \left|\frac{A - 1}{A + 1}\right|$$

defines an ellipse in the complex z-plane with foci at $z = 0$ and $z = 1$, and passing through the point $z = a$, see Figure D.2. Moreover, the inequality

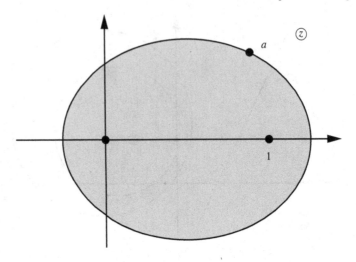

Fig. D.2 The ellipse in Lemma D.1.

$$\left|\frac{Z-1}{Z+1}\right| > \left|\frac{A-1}{A+1}\right|$$

defines the interior of the ellipse containing the origin in the z-plane, and the inequality

$$\left|\frac{Z-1}{Z+1}\right| < \left|\frac{A-1}{A+1}\right|$$

defines the domain exterior to the ellipse in the z-plane.

Proof. The real constant

$$C = \left|\frac{A-1}{A+1}\right| < 1$$

since the complex number $A = \left(1 - a^{-1}\right)^{1/2}$ is located in the right-hand side of the complex A-plane, i.e., $\operatorname{Re} A > 0$. For given A, the complex numbers Z satisfying

$$|Z-1| = C|Z+1|$$

define a circle in the complex Z-plane with center at Z_0 and radius R, where, see Section D.1.1,

$$\begin{cases} Z_0 = \dfrac{1+C^2}{1-C^2} \\[2mm] R = \dfrac{2C}{|1-C^2|} \end{cases}$$

and the circle encircles the point $Z = 1$.

The parameter representation of the circle

$$Z = Z_0 + Re^{i\phi}, \qquad \phi \in [0, 2\pi)$$

implies that the curve in the z-plane is

$$z = \frac{1}{1 - Z^2} = \frac{1}{(1 + Z)(1 - Z)} = \frac{1}{(1 + Z_0 + Re^{i\phi})(1 - Z_0 - Re^{i\phi})}$$

However, this is an ellipse with foci at $z = 0$ and $z = 1$, since

$$\begin{cases} |z| = \dfrac{1}{|1 + Z||1 - Z|} = \dfrac{1}{C|1 + Z|^2} \\[2mm] |z - 1| = \dfrac{|Z|^2}{|1 + Z||1 - Z|} = \dfrac{|Z|^2}{C|1 + Z|^2} \end{cases}$$

and

$$\begin{cases} |Z|^2 = (Z_0 + R\cos\phi)^2 + R^2 \sin^2\phi = \dfrac{(1 + C^2 + 2C\cos\phi)^2 + 4C^2 \sin^2\phi}{(1 - C^2)^2} \\[2mm] |1 + Z|^2 = (Z_0 + 1 + R\cos\phi)^2 + R^2 \sin^2\phi = 4\dfrac{(1 + C\cos\phi)^2 + C^2 \sin^2\phi}{(1 - C^2)^2} \end{cases}$$

We simplify

$$\begin{cases} |Z|^2 = \dfrac{(1 + C^2)^2 + 4C(1 + C^2)\cos\phi + 4C^2}{(1 - C^2)^2} \\[2mm] |1 + Z|^2 = 4\dfrac{1 + 2C\cos\phi + C^2}{(1 - C^2)^2} \end{cases}$$

The sum of the distances from z to the origin and 1 is

$$|z| + |z - 1| = \frac{(1 - C^2)^2 + (1 + C^2)^2 + 4C(1 + C^2)\cos\phi + 4C^2}{4C(1 + 2C\cos\phi + C^2)}$$

$$= \frac{2(1 + C^2)^2 + 4C(1 + C^2)\cos\phi}{4C(1 + 2C\cos\phi + C^2)} = \frac{1 + C^2}{2C}$$

which is the equation of the ellipse with foci at $z = 0$ and $z = 1$. Notice that

$$\frac{1 + C^2}{2C} = \frac{|A + 1|^2 + |A - 1|^2}{2|A + 1||A - 1|} = \frac{|A|^2 + 1}{|A^2 - 1|} = \frac{|1 - 1/a| + 1}{1/|a|} = |a| + |a - 1|$$

by the parallelogram law $|z_1 + z_2|^2 + |z_1 - z_2|^2 = 2|z_1|^2 + 2|z_2|^2$, and we have proved that the

$$|Z - 1| = C|Z + 1|$$

defines an ellipse in the complex z-plane.

The complex number $z = 1$ lies inside the ellipse, and it corresponds to $Z = (1 - z^{-1})^{1/2} = 0$, and the complex number $Z = 0$ satisfies

$$|Z - 1| > C|Z + 1|$$

since

$$C < 1$$

The inequality

$$|Z - 1| > C|Z + 1|$$

therefore defines the domain inside the ellipse, and

$$|Z - 1| < C|Z + 1|$$

the outside. □

Appendix E
Elementary and special functions

A long list of elementary and special functions can be expressed in the hypergeometric function $_2F_1(\alpha,\beta;\gamma;z)$, and its two confluent versions, $_1F_1(\alpha;\gamma;z)$ and $_0F_1(\gamma;z)$, respectively. Some examples are given in this appendix. The Greek letters α and β are arbitrary complex numbers, and n and m are non-negative integers.

E.1 Hypergeometric function $_2F_1(\alpha,\beta;\gamma;z)$

Elementary functions

$$\begin{cases} (1+z)^\alpha = F(-\alpha,\beta;\beta;-z) \\ \ln(1+z) = zF(1,1;2;-z) \\ \arctan z = zF\left(\frac{1}{2},1;\frac{3}{2};-z^2\right) \\ \arcsin z = zF\left(\frac{1}{2},\frac{1}{2};\frac{3}{2};z^2\right) \end{cases}$$

Elliptic integrals

Complete elliptic integral of the first kind

$$K(m) = \int_0^1 \left((1-t^2)(1-mt^2)\right)^{-1/2} dt = \int_0^{\pi/2} \left(1-m\sin^2\theta\right)^{-1/2} d\theta$$
$$= \frac{\pi}{2}F\left(\frac{1}{2},\frac{1}{2};1;m\right)$$

Complete elliptic integral of the second kind

$$E(m) = \int_0^1 \left(1 - t^2\right)^{-1/2} \left(1 - mt^2\right)^{1/2} dt = \int_0^{\pi/2} \left(1 - m\sin^2\theta\right)^{1/2} d\theta$$
$$= \frac{\pi}{2} F\left(-\frac{1}{2}, \frac{1}{2}; 1; m\right)$$

Jacobi polynomials

$$P_n^{(\alpha,\beta)}(x) = \binom{n+\alpha}{n} F\left(-n, n+\alpha+\beta+1; \alpha+1; \frac{1-x}{2}\right)$$

Legendre functions

Legendre functions of the first kind

$$P_\nu(z) = F\left(-\nu, \nu+1; 1; \frac{1-z}{2}\right)$$

Legendre functions of the second kind

$$Q_\nu(z) = \frac{\sqrt{\pi}}{(2z)^{\nu+1}} \frac{\Gamma(\nu+1)}{\Gamma(\nu+3/2)} F\left(1+\frac{\nu}{2}, \frac{1+\nu}{2}; \nu+\frac{3}{2}; \frac{1}{z^2}\right)$$

Associated Legendre functions

$$\begin{cases} P_\nu^m(x) = (1-x^2)^{\frac{m}{2}} \dfrac{d^m}{dx^m} P_\nu(x) \\ Q_\nu^m(x) = (1-x^2)^{\frac{m}{2}} \dfrac{d^m}{dx^m} Q_\nu(x) \end{cases} \quad x \in [-1,1]$$

$$\begin{cases} \mathscr{P}_\nu^m(z) = (z^2-1)^{\frac{m}{2}} \dfrac{d^m}{dz^m} P_\nu(z) \\ \mathscr{Q}_\nu^m(z) = (z^2-1)^{\frac{m}{2}} \dfrac{d^m}{dz^m} Q_\nu(z) \end{cases} \quad z \in \mathbb{C}$$

Tchebysheff polynomials

Tchebysheff first kind

$$T_n(\cos\theta) = \cos n\theta = F\left(-n,n;\frac{1}{2};\frac{1-x}{2}\right)$$

Tchebysheff second kind

$$U_n(\cos\theta) = \frac{\sin(n+1)\theta}{\sin\theta} = (n+1)F\left(-n,n+2;\frac{3}{2};\frac{1-x}{2}\right)$$

E.2 Confluent functions $_1F_1(\alpha;\gamma;z)$

Elementary functions

$$\begin{cases} e^z = {}_1F_1(\alpha;\alpha;z) \\ \dfrac{e^{-iz}}{z}\sin z = {}_1F_1(1;2;-2iz) \\ \dfrac{e^z}{z}\sinh z = {}_1F_1(1;2;2z) \end{cases}$$

Bessel functions

$$J_\nu(z) = \frac{z^\nu e^{-iz}{}_1F_1(\nu+1/2;2\nu+1;2iz)}{2^\nu\Gamma(\nu+1)}$$

Laguerre polynomials

$$L_n^{(\alpha)}(z) = \binom{n+\alpha}{n}{}_1F_1(-n;\alpha+1;z)$$

Hermite polynomials

$$\begin{cases} H_{2m}(x) = (-1)^m 2^{2m} m! L_m^{(-1/2)}(x^2) \\ H_{2m+1}(x) = (-1)^m 2^{2m+1} m! x L_m^{(1/2)}(x^2) \end{cases} \quad x \in \mathbb{R}$$

E.2.1 Error functions

$$\mathrm{erf}(z) = \frac{2z}{\sqrt{\pi}} \, {}_1F_1(1/2; 3/2; -z^2) = \frac{2z}{\sqrt{\pi}} e^{-z^2} \, {}_1F_1(1; 3/2; z^2)$$

E.3 Confluent functions ${}_0F_1(\gamma; z)$

Bessel functions

$$J_\nu(z) = z^\nu \frac{{}_0F_1(\nu+1; -z^2/4)}{2^\nu \Gamma(\nu+1)}$$

Appendix F
Notation

Most of the notation and symbols adopted in this textbook are traditional, and there is very little risk of confusion, but for the sake of completeness, we collect the symbols in this appendix.

- We use the notation $\mathbb{Z}$ for all integers $0, \pm 1, \pm 2, \dots$.
- The positive integers $1, 2, 3, 4, \dots$ are denoted $\mathbb{Z}_+$.
- The negative integers $-1, -2, -3, -4, \dots$ are denoted $\mathbb{Z}_-$.
- The natural, non-negative, integers $0, 1, 2, 3, \dots$ are denoted $\mathbb{N}$.
- The field of real numbers is denoted $\mathbb{R}$.
- The field of complex numbers is denoted $\mathbb{C}$. Sometimes the point at infinity is included, and it is then appropriate to view the field as the Riemann sphere.
- We use the symbols o and O defined by

$$
\begin{cases}
f(x) = o\left(g(x)\right), & x \to a \quad \Longleftrightarrow \quad \lim\limits_{x \to a} \dfrac{f(x)}{g(x)} = 0 \\[2ex]
f(x) = O\left(g(x)\right), & x \to a \quad \Longleftrightarrow \quad \dfrac{f(x)}{g(x)} \text{ bounded in a neighborhood of } a
\end{cases}
$$

- The symbol $\square$ is used to end a proof.
- The symbol $\blacksquare$ is used to end an example or a comment.
- The dagger, †, in front of a problem denotes a more difficult problem.
- The star, *, denotes complex conjugation of a complex number.

References

1. Abramowitz, M., Stegun, I.A. (eds.): Handbook of Mathematical Functions. Applied Mathematics Series No. 55. National Bureau of Standards, Washington, D.C. (1970)
2. Andrews, G., Askey, R., Roy, R.: Special Functions. Volume 71 of Encyclopedia of Mathematics and its Applications. Cambridge University Press, Cambridge, U.K. (1999)
3. Arfken, G.B., Weber, H.J.: Mathematical Methods for Physicists. Academic Press, New York (1995)
4. Carlson, B.C.: Special Functions of Applied Mathematics. Academic Press, New York (1977)
5. Eisenhart, L.P.: Separable systems in Euclidean 3-space. Phys. Rev. **45**(6), 427–428 (1934)
6. Elaydi, S.: An Introduction to Difference Equations, third edn. Springer-Verlag, New York (2005)
7. Erdélyi, A.: Certain expansions of solutions of the Heun equation. Quart. J. Math. (Oxford) **15**, 62–69 (1944)
8. Erdélyi, A., Magnus, W., Oberhettinger, F., Tricomi, F. (eds.): Higher Transcendental Functions. 3 vols. Bateman Manuscript Project. McGraw-Hill, New York (1953)
9. Greene, R.E., Krantz, S.G.: Function Theory of One Complex Variable, third edn. American Mathematical Society, Providence, R.I. (2006)
10. Henrici, P.: Applied and Computational Complex Analysis, vol. 1. John Wiley & Sons, New York (1974)
11. Henrici, P.: Applied and Computational Complex Analysis, vol. 2. John Wiley & Sons, New York (1977)
12. Hille, E.: Ordinary Differential Equations in the Complex Domain. Dover Publications, Mineola, N.Y. (1976)
13. Hille, E.: Analytic Function Theory, vol. 1, second edn. Chelsea Publishing Company, New York (1982)
14. Hochstadt, H.: The Functions of Mathematical Physics. John Wiley & Sons, New York (1971)
15. Magnus, W., Oberhettinger, F., Soni, R.P.: Formulas and Theorems for the Special Functions of Mathematical Physics. Springer-Verlag, New York (1966)
16. Maier, R.: The 192 solutions of the Heun equation. Math. Computat. **76**(258), 811 (2007)
17. Miller Jr., W.: Lie Theory and Special Functions. Academic Press, New York (1968)
18. Morse, P.M., Feshbach, H.: Methods of Theoretical Physics, vol. 1. McGraw-Hill, New York (1953)
19. Paris, R., Kaminski, D.: Asymptotics and Mellin–Barnes Integrals. Cambridge University Press, Cambridge, U.K. (2001)
20. Rainville, E.: Special Functions. Chelsea Publishing Company, New York (1960)
21. Ronveaux, A.: Heun's Differential Equations. Oxford University Press, Oxford (1995)
22. Rudin, W.: Principles of Mathematical Analysis. McGraw-Hill, New York (1976)
23. Slavyanov, S.Y., Lay, W.: Special Functions: A Unified Theory Based on Singularities. Oxford University Press, Oxford (2000)

24. Stromberg, K.R.: An Introduction to Classical Real Analysis. Wadsworth International Group, Belmont (1981)
25. Svartholm, N.: Die Lösung der Fuchsschen Differentialgleichung zweiter Ordnung durch hypergeometrische Polynome. Mathematische Annalen **116**(3), 413–421 (1939)
26. Szegö, G.: Orthogonal Polynomials, fourth edn. American Mathematical Society, Providence, R.I. (1985)
27. Talman, J.D.: Special Functions, A Group Theoretic Approach. W. A. Benjamin, Inc., New York (1968)
28. Temme, N.M.: Special Functions: An Introduction to the Classical Functions of Mathematical Physics. Wiley-Interscience, New York (1996)
29. Vilenkin, N.: Special Functions and the Theory of Group Representations. American Mathematical Society, Providence, R.I. (1968)
30. Watson, G.N.: A Treatise on the Theory of Bessel Functions, second edn. Cambridge University Press, Cambridge, U.K. (1966)
31. Whittaker, E.T., Watson, G.N.: A Course of Modern Analysis, fourth edn. Cambridge University Press, Cambridge, U.K. (1969)

Index